BOTANICAL RESEARCH AND PRACTICES

AF825204

A MONOGRAPH ON *SOLANUM TORVUM* SWARTZ

BOTANICAL RESEARCH AND PRACTICES

Additional books in this series can be found on Nova's website under the Series tab.

Additional e-books in this series can be found on Nova's website under the e-book tab.

BOTANICAL RESEARCH AND PRACTICES

A MONOGRAPH ON *SOLANUM TORVUM* SWARTZ

ZUBAIDA YOUSAF
ANA AKRAM
AND
HAFIZA AYESHA REHMAN
EDITORS

NOVINKA

New York

NOTICE TO THE READER

LIBRARY OF CONGRESS CATALOGING-IN-PUBLICATION DATA

ISBN: 978-1-62808-421-4

Published by Nova Science Publishers, Inc. † New York

CONTENTS

PREFACE

We all strongly believe that life develops under the canopy of plants. Basic needs of human beings (Food, Shelter and Medicine) are fulfilled from the plant resources. From the time of Adam, human beings have treated their diseases with various medicines. Plants have played an important role in human lives from the days of Adam and Eve.

Plants have been used as a source of food, nutrition, shelter, and medicines. In ancient Egypt circa 1550 and in Babylon circa 1770 BC, plants have been in use for medicinal purposes. The recovery of medicinal plants from Giza pyramids have proved the importance of medicinal plants in that area when the ancient Egyptians have the common religious practice of burying medicinal plants with their dead. They believed that the dead will live again and will utilize these medicinal plants. It is estimated that 25,000-75,000 plants species are used by the people for curing various diseases, only 1% of this is known to scientists and are utilized for commercial purposes. Those plants which are used in indigenous herbal medicines but not in practice in modern medicines and commercial products may be categorized in underutilized species.

Humanity relies on a diverse range of cultivated species; at least 6000 such species are used for a variety of purposes. It is often stated that only a few staple crops produce the majority of the food supply. This might be correct but the important contributions of many minor species should not be underestimated. Agricultural research has traditionally focused on these staples; while relatively little attention has been given to minor (or underutilized or neglected) crops, particularly by scientists in developed countries. Such crops have, therefore, generally failed to attract significant research funding. Unlike most staples, many of these neglected species are adapted to various marginal growing conditions such as those of the Andean and Himalayan highlands, arid areas, salt-affected soils, etc.

Furthermore, many crops considered to be neglected at a global level are staples at a national or regional level (e.g. tuff, fonio, Andean roots and tubers, etc.), contribute considerably to food supply in certain periods (e.g. indigenous fruit trees) or are important for a nutritionally well-balanced diet (e.g. indigenous vegetables). The limited information available on many important and frequently basic aspects of neglected and underutilized crops hinders their development and their sustainable conservation. One major factor hampering this development is that the information available on germplasm is scattered and not readily accessible, i.e. only found in 'grey literature' or written in minority languages. Moreover, existing knowledge on the genetic potential of neglected crops is limited. This has frequently resulted in uncoordinated research efforts for most neglected crops, as well as inefficient approaches to the conservation of these genetic resources.

This monograph intends to draw attention to *Solanum torvum* Swartz, which have been neglected in varying degrees by researchers or have been underutilized economically? It is hoped that the information compiled will contribute to:

(1) Identifying constraints and possible solutions to the use of the *Solanum torvum* swartz.
(2) Detecting existing gaps in available conservation and use approaches.

This monograph intends to contribute to improvement of the potential value of *Solanum torvum* Swartz, through the increased use of the available genetic diversity In addition, it is hoped that the monographs in the series will form a valuable reference source for all those scientists involved in conservation, research, improvement and promotion of this plant species.

Zubaida Yousaf
Ana Akaram
Ayesha Rehman

ABSTRACT

Solanum torvum Swartz. is a small Shrub of family Solanaceae, distributed widely in Pakistan, India, Malaya, China, Philippines, and tropical America. For many decades, different ethnic groups have used the dried stem and root of this plant for treatment of various ailments. Number of saponins, steroidal alkaloids has been isolated from this plant. Nine very commonly known compounds including neochlorogenin 6-*O*-*β*-D-quinovopyranoside (**1**), neochlorogenin 6-*O*-*β*-D-xylopyranosyl-(1→3)-*β* D-quinovopyranoside (**2**), neochlorogenin 6-*O*-*α*-L-rhamnopyranosyl-(1→3)-*β*-D-quinovopyranoside (**3**), solagenin 6-*O*-*β*-D quinovopyranoside (**4**), solagenin 6-*O*-*α*-L-rhamnopyranosyl-(1→3)-*β*-D-quinovopyranoside (**5**), isoquercetin (**6**), rutin (**7**), kaempferol (**8**) and quercetin (**9**) were isolated from *Solanum torvum* first time by silica gel, Sephadex LH-20, and Rp-C18 column chromatography. Two solanolide 6-*O*-[-l-rhamnopyranosyl-(1→3)-*O*-d-quinovopyranoside] (**1**), solanolide 6-*O*-[_-d-xylopyranosyl-(1→3)-*O*-d-quinovopyranoside] (**2**), yamogenin 3-*O*-[-d-glucopyranosyl-(1→6)-*O*-d-glucopyranoside] (**3**) and neochlorogenin 3-*O*-[-dglucopyranosyl-(1→6)-*O*-d-glucopyranoside] (**4**) on the basis of spectroscopic analysis. They were found effective against human cancer cells. The C-4 sulfated isoflavonoid, torvanol A (1), and the steroidal glycoside, torvoside H (3), together with the known glycoside, torvoside A (2), were isolated from a methanolic extract of *Solanum torvum*fruits, reported to have antiviral activity.

Polygalacturonase-inhibiting protein was isolated from *Solanum torvum* by RT-PCR and RACE, designated StPGIP. The cDNA sequence of StPGIP was 1097 bp long and contained an open reading frame of 990bp. Polygalacturonase-inhibiting proteins. (PGIPs) have been found to help in preventing the invasion of fungus including V. dahlia that cause verticillum wilt disease in egg plant. The methanolic extract of *Solanum torvum* fruit showed a wide spectrum of antimicrobial activities against human and animal clinical isolates. *Solanum torvum* fruit containing phenolic compounds has great potential as a natural source of anti-diabetic and antioxidant drugs and resistance against fungal diseases and bacterial wilt.

Ethanolic extract from *Solanum torvum* could prevent development of high blood pressure induced by a diet rich in fructose probably by reversing the metabolic alterations induced by fructose. It also causes neurological and steroidal glycosides diseases.

Oral chronic administration of AEST induced potentiation of arterial hypertension and cardiac hypertrophy in l-NAME treated rats. These effects may result from a reduction in sensitivity to vasorelaxant agents and increase in hypersensitivity to contractile e factors. AEST possess potent *in vitro* vasocontractile activity that may result from activation of both 1-adrenergic pathway and calcium influx. Cadmium (Cd) concentration in eggplant (*Solanum melongena*) fruits can be drastically reduced by grafting them with *Solanum torvum* rootstock.

INTRODUCTION

The family Solanaceae is among the most economically and medicinally important families of angiosperms. The genus *Solanum* is a hyper-diverse taxon of this family. There are about 2000 species of *Solanum* in the world that are mainly distributed in the tropical and sub-tropical areas, with a small number in the temperate areas (Jennifer and James, 1997). About 21 species and one variety in this genus are used as herbal medicines. *Solanum torvum* Swartz. is a small Solanaceous shrub, distributed widely in Pakistan, India, Malaya, China, Philippines, and tropical America (Nasir, 1985). For many decades, different ethnic groups have used the dried stem and root of this plant for treatment of various ailments. Its Chinese medicinal name is Jinniukou (Annonymous, 2000). Among the major chemical constituents of *S. torvum* are steroids, steroid saponins, steroid alkaloids, and phenols. Pharmacological studies indicate that the stem and root of *S. torvum* have anti-tumor, anti-bacterial, anti-viral, anti-inflammatory, and other medicinally important effects.

I. TAXONOMICAL CLASSIFICATION

Kingdom:	Plantae, Plants
Subkingdom:	Tracheobionta, (Vascular plants)
Super division:	Spermatophyta, (Seeds plants
Division:	Angiosperma
Class:	Dicotyledons
Order:	Tubiflorae
Family:	Solanaceae
Genus:	*Solanum*
Species:	*torvum.*

Chapter 1

BOTANY OF *SOLANUM TORVUM* SWARTZ

Solanum torvum Swartz.Nov. Gen & Sp. Pl. 47.1788; Wight, Pl. Ind. Or. 2: t. 345; Clarke in Hook. f., Fl. Brit. Ind. 4: 234. 1883; R. R. Stewart, an Anno. Catalogue of vascular plants from West Pakistan and Kashmir In: Nasir, E. and S.I. Ali. Flora of West Pakistan. 645; Shang, *et. al.*, flora China. 329. 1975;Nasir, Fl. Pak. 168: 10. 1985.

I. SYNONYMY

Solanum ficifolium auctt. Non Jacquem.: Roxb., Fl. Ind. (Car.ed) 2, 1: 572. 1832.

II. COMMON NAME

Pea eggplant, cherry eggplant, devil's fig, plate brush, Turkey berry (En). Susumber berrym (En),Susumba, Soushumber, Mélongène-diable, bellangère bâtarde, aubergine pois (Fr).

III. VERNACULAR NAMES

Sundai, Shui qie (chinese), Teufels-Nachtschatten (Germen), Takokak, Pokak (Indoneshia), Terong pipit (Malay), Jurubeba (Brazil), Berenjena cimarrona (Spanish), Terongan (Spainsh), Artaubergin (Swedish).

Shrub, 100-300cm tall. Stem and branches are sparsely prickly, stellate tomentose. Stem with stout, reversed, reddish or pale yellow prickles, 0.2-0.1 × 0.2-0.1cm, sometime basal stellate hairs. Leaf petiole 2-4cm long, Leaves solitary or paired, ovate-sinuate, many branch of stellate hairs on above surface, lobes rarely deep, never prickly 9-13 cm long, 5-10cm wide, margin sinueate or usually 5-7 lobed, apex acute. Inflorescence extra axillary many flowered racemose panicle cymes, peduncle mostly 1 or 2 branched, short, 0.1-0.4cm long, pubescent. Flower andro monoecious, pale white, pedicel dark slender, 0.5-0.1cm long, bearing simple glandular and stellate hairs. Sepals 0.2-0.5cm long, lanceolate, sparingly hairy. Corolla glabrous, 0.3-0.5cm long, stellate pubescence abaxially. Filament 0.1cm long, anthers 0.4-0.7cm long ovary and style glabrous, 0.6-0.8cm long. Berry yellow, smooth, 1-1.5cm long, calyx lobes present. Seeds, discoid, 0.2cm broad, smooth, yellowish brown in color.

IV. MORPHOLOGICAL CHARACTERS OF *SOLANUM TORVUM* SWARTZ

(Morphological characters of the species studied from the available herbaria (National Herbarium, NARC; Karachi University Herbarium, Karachi; Quaid-e-Azam University herbarium, Islamabad; Herbarium of Pakistan Museum of Natural History, Islamabad) specimens of S. torvum)

Flowering Period: April-May
Fruiting Period: it continues after the shrubs have reached about 1 to 1.5 m
Type: West Indies, Swartz (S).

V. ORIGIN

Origin: Tropical America

VI. Distribution and Status

Distribution:	Native to West Indies, India, Myanmar, Thailand, Philippines, Malaysia, China and trotatus: Frequentpical America. Widely naturalised in South and South East Asia.
Status:	Frequent

Chapter 2

TAXONOMIC SUMMARY OF *SOLANUM* L. SECTION *SOLANUM*

The section *Solanum,* centering on the species commonly known as the black, garden or common nightshade, *Solanum nigrum* L., is one of the largest and most variable species groups of the genus. It more usually referred to as the section *Morella (Maurella* (Dun.) Dumort.; *Morella* (Dun.) Bitt.), but since the generic type is *S. nigrum,* the correct name for this section is *Solanum* (Seithe 1962). Species belonging to this section were distributed from temperate to tropical regions, and from sea level to altitudes over 3500 meters. Though, these species groupswere often referred to as the *Solanum nigrum* complex, the section was composed of a large number of morpho-genetically distinct taxa, which shows their greatest diversity and concentration in the New World tropics, 10 Black nightshades, *Solanum nigrum* L. and related species particularly in South America. *Solanum nigrum* itself is a predominantly Eurasian species, which does not occur naturally in South America (Edmonds 1979a). In most parts of the world, particularly in Europe and North America, these species are considered troublesome weeds in agriculture, but in many developing countries they constitute a minor food crop, with the shoots and berries not only being used as vegetables and fruits, but also for various medicinal and local uses.

The genus *Solanum* is hyper diverse genus of the family of Solanaceae. Phenotypic complexity of the genus contributes toward the taxonomic confusions of various species. The taxonomic complexity species associated with the section *Solanum* have long been accepted (Stebbins and Paddock 1949; Heiser 1963; Symon 1970; Venkateswarlu and Rao 1972; D'Arcy 1973, 1974). Various classical, experimental and numerical studies demonstrated that the complexity is attributable to a number of causes (Edmonds 1972, 1977, 1979a). Among these are the following.

Species belonging to this section have been subjected to extensive taxonomic study since Dillenius first delimited four taxa with polynomials in 1732. Linnaeus subsequently modified Dillenius' work, describing, six varieties under the binomial *S. nigrum* in 1753. Since then, the species related to *S. nigrum* have been reclassifiedseveral times. Characters used by later taxonomists to, separate, and describe additional taxa often differ very slightly from those given for species by earlier workers. Over 300 post-Linnaean specific and infraspecific names have now been published, and synonymy is extensive within the section. (Edmonds 1977). However, no satisfactory revision of the whole section has yet been devised. The boundaries between many of the species are still very difficult to define, with many of the 'new' taxa proving to be no more than slight morphological variants of those already described. The situation has been further complicated by a number of authors who have persistently treated different members of the section as belonging to one highly variable species, namely *S. nigrum* (Tandon and Rao 1966a, 1966b, 1974; Rao and Tandon 1969; Rao *et al.,* 1971, 1977; Venkateswarlu and Rao 1971, 1972). Fortunately, Ganapathi and Rao (1986) subsequently demonstrated that the Indian plants generally described as 'cytotypes' of *S. nigrum* were indeed conspecific with the distinct diploid tetraploid and hexaploid species recognized by western workers, so it is hoped that the correct taxonomic names will now be adopted for the taxa concerned.

I. Phenotypic Plasticity

Solanum species (particularly S. torvum and its wild relatives) display varying amounts of phenotypic variation, particularly in their vegetative features such as plant habit, leaf size and form.. In addition, senescence is often accompanied by smaller and fewer flowers and fruits than usual, while the gene for anthocyanin pigmentation in flowers seems to be dependent on light intensity and temperature for its expression, in some species (Baylis 1958; Henderson 1974; Edmonds 1977).

II. POLYPLOIDY

Species belonging to the section Leptostemon constitute a polyploid series, with diploid (2*n*=2*x*=24), tetraploid (2*n*=4*x*=48) and hexaploid (2*n*=6*x*=72) species occurring throughout most of the geographical range in which the section is found. Octoploid plants (2*n*=8*x*=96) also have been reported on two occasions (Heiser 1963; Edmonds 1977, 1979a). Edmonds has made somatic chromosome counts of over 200 accessions of known origin collected throughout the world, and subsequently used experimentally, though not all of them were published (Edmonds1972, 1977, 1982, 1983, 1984a, 1986). Many other authors have reported counts for small species groups, but some of these were considered based on misidentified material. The possibility of determining a rough estimate of ploidy level from herbarium material was discussed in Edmonds (1979a). Both pollen diameter and stomata length tend to increase with ploidy level, though pollen sizes are not always directly proportional to chromosome number in these species (Heiser 1963; Edmonds 1979a), if used in association with stomatal length, a good estimate of ploidy level usually can be obtained.

III. NATURAL HYBRIDIZATION

Species belonging to the section *Solanum* were predominantly self-pollinating, but outside cross-breeding can occur, and natural inter- and intra-specific hybridizations, especially among the smaller-flowered diploids, have been reported by a number of workers (e.g. Stebbins and Paddock 1949; D'Arcy 1974; Henderson 1974). Natural hybrids also have been reported at higher ploidy levels, e.g. infra-specific hybrids of the hexaploid *S. nigrum* (Venkateswarlu and Rao 1972). In addition, spontaneous hybrids have been reported between taxa of different ploidy levels, as between the hexaploid *S. scabrum* and the diploid *S. americanum* (Henderson 1974) and between 12 Black night shades. *Solanum nigrum* L., and related species the hexaploid *S. nigrum* and the diploid known as *S. physalifolium* Rusby var. *nitidibaccatum* (Bitt.) Edmonds (Leslie 1978; Edmonds 1981).

Natural hybridization is probably more widespread in this section than generally supposed. Though this was probably followed by subsequent genetic breakdown in F1 or F2 generations (Edmonds 1977). It may also be followed by back-crossing to the parental species. This would result in morpho-genetically complex population variation: the collection of specimens from such populations would explain some of the difficulties encountered in the morphological differentiation of these species in the herbarium (Edmonds 1979a).

IV. Conflicting Variation

Jardine and Sibson (1971) defined discordant variation as "A pattern of variation in which the relative extents to which populations are differentiated do not coincide with different sets of attributes. The product of mosaic evolution. A population study investigating the degree of discordance in 32 *Solanum* populations representing 20 species of four differing ploidy levels demonstrates that the taxa exhibit considerable inherent discordant variation, and that a large number of characters must be used for the differentiation of these species (Jardine and Edmonds 1974; Edmonds 1978b).

V. Identification of Species

Various taxonomic treatments of species belonging to the section *Solanum*were fully discussed in Edmonds (1977, 1979a). Earlier workers were essentially 'splitters' based on large numbers of inter and intraspecific taxa on characters now known to be insignificant in this variable species group. Most of the more recent authors, however, based their conclusions on material of fully documented, known wild origin and, in many cases, on considerable experimental work. As a result, the taxonomy of the species involved has been resolved for certain geographical areas, where the species relationships with the component taxa are now fairly well understood.

VI. Ecology

In Puerto Rico, turkey berry that grows in upland sites have received from about 1000 to 4000 mm of annual precipitation. It also grows in riparian zones in drier areas. Turkey berry grows on all types of moist, fertile soil at elevations from near sea level to almost 1,000 m in Puerto Rico (Little and others 1974) and 2,000 m in Papua New Guinea (Pacific Island Ecosystems at

Risk 2001). Given an equal start after disturbance, turkey berry quickly top most herbs, grasses, and other shrubs. It grows best in full sunlight and does well in light shade or shade for part of the day, but cannot survive under a closed forest canopy. Turkey berry single plants, groups, and thickets are most frequently seen on roadsides, vacant lots, brushy pastures, recently abandoned farmland, landslides, and river banks.

Chapter 3

REPRODUCTION

Flowering and fruiting were continued after the shrubs got to about 1 to 1.5 m in height. Ripe fruits collected in Puerto Rico averaged 1.308 + 0.052 g. Air dry seeds from these fruits weighed an average of 0.00935 g or 1,070,000 seeds/kg. These seeds were sown on commercial potting mix and 60 percent germinated between 13 and 106 days after sowing. The seedlings were common in recently disturbed ground. Frugivorous birds eat the fruits and spread the seeds (Pacific Island Ecosystems at Risk 2001). Nevertheless, Turkey berry can be propagated vegetatively by placing branch cuttings, with or without leaves, in a mist chamber for one month (Badola and others 1993).

I. GROWTH AND MANAGEMENT

Turkey berry grows about 0.75 to 1.5 m in height per year Their life span is relatively short; as most of them live for a period of 2 years. Physical control of the shrub may be done by grubbing out the plants. However, they can be killed by translocate herbicides applied to the leaves or the cut stumps (Pacific Island Ecosystems at Risk 2001).

II. BENEFITS

The wood is soft, light and of little use except for emergency fuel. The fruits are edible and used in Thai cuisine (Royal Horticultural Society 2001) as well as incorporated into soups and sauces in the Ivory Coast (Herzog and

Gautier-Béguin 2001. Turkey berry contains a number of potentially active chemicals including the sapogenin steroid and chlorogenin (Badola and others 1993). Aqueous extracts of turkey berry are lethal to mice or depress the erythrocytes, leukocytes and platelets in their blood (Tapia and others 1996). A related chemical, cholecalciferol, is the active ingredient in a number of commercial rodentacides (American Board of Veterinary Toxicology 2001). Extracts of the plant are reported to be useful in the treatment of hyperactivity (Null 2001), colds and cough (CPR Environmental Education Centre 2001), pimples, skin diseases, and leprosy (Liogier 1990). Turkey berry is being crossed with eggplant in an attempt to incorporate genes for resistance to *Verticillium* wilt into the vegetable (Bletsos and others 2001).

Chapter 4

TRADITIONAL MEDICINAL USES

Solanum torvum is a pharmacologically important species from the family of Solanaceae. Traditional medicinal uses of *S. torvum*, have been highlighted in the Ayurveda and Chinese pharmacopeia. These traditional uses are summarized in the following table (Table 1).

Table 1. Traditional uses of *Solanum torvum*

Traditional uses	Method of utilization	Part used	Community using *S. torvum*	Reference
Relief from colds andcoughs.	Powder mixed with hot water or cow milk and administrated orally	Leaves(after drying in shade)	Puducherry, Karaikal, Mahe andYanam, India	Udayakumar et al., 2009
To cure cracked footCuring coughs	Externally applied on cracks.Powdered fried fruit is taken for cough	Root (Extraction)Fruit(after frying)	Kurichyas, Kannur district	Rajith& Ramachandran, 2010
To reduce body heat	Leaf juice take orally	Leaf (extractions)	Kancheepuram	Chellaiah et al., 2006
Lethal to mice	Aqueous extract of berry	Fruit (extraction)	Mexico	Tapia et al., 1996
Edible fruit	Cooked fruit as an important ingredient of soups and sauces	Fruit	India	Badola, 1993

Table 1. (Continued)

Traditional uses	Method of utilization	Part used	Community using *S. torvum*	Reference
To control bacterial and fungal diseases of *S. melongena*	Intercropping cultivation	Whole plant	Indonesia	Gousset, et al., 2005
Asthma, Diabetes, hypertension	Combination of root and leaf juice	Root and leaves	Garo Tribe Bangladesh	Rahmatullah, et al., 2009
Vermifuge	Cooked fruit	Fruit	Tirunelveli, Tamil Nadu, India	Mohan, et al, 2009
To treat liver diseases, tuberculosis, and as anti-anemic	Root Juice	Roots	Northeastern Brazile	Agra, 1999

Chapter 5

PHYTOCHEMISTRY

Solanum torvum has been extensively explored for its chemical constituents. Various parts (fruit, leaves and roots) were being used for the isolation of a wide of compounds. This plant species is a very good source of alkaloids, flavonoids, saponins, tannins, and glycosides (Chah *et al.,* 2000). According to Pérez-Amador *et al.,* (2007), the percentage composition of various compounds within this species is total alkaloid content (0.12%) total glycoalkaloids (0.038 %), and glycosylated compounds derived from solasodine, namely solasonine (0.0043%) and solamargine (0.0028%). Kusirisin *et al.,* in 2009 recorded polyphenolic compounds that included phenol, flavonoids and tannin. The concentrations of these compounds were recorded as 160.30, 104.36 and 65.91 mg/g, respectively.

I. COMPOUNDS ISOLATED FROM FRUIT

Antiviral isoflavonoid sulfate and steroidal glycosides were also isolated from the fruits of *S. torvum*. Arthan *et al.,* 2002 investigatedMeOH extracts of fruit and found one new isoflavonoid sulfate named as torvanol A, and a new steroidal glycoside, named torvoside H, together with the already-known glycoside, torvoside A.

II. Compounds Isolated from Aerial Parts

Aerial parts of *S. torvum* are rich sources of steroid and saponins. Lu *et al.,*1983 isolated four steroidal compoundssolanolide 6-O-[α-l-rhamnopyranosyl-(1→3)-O-β-d-quinovopyranoside], solanolide 6-*O*-[β-d-xylopyranosyl-(1 → 3)-*O*-β-d-quinovopyranoside], yamogenin 3-*O*-[β-D-glucopyranosyl-(1 → 6)-*O*-β-D-glucopyranoside] and solanolide 6-O- [α-l-rhamnopyranosyl-(1 → 3)-O-β-d-quinovopyranoside]. Yuan-Yuan *et al.,* 2011 reported the isolation of two novelty C-22 steroidal lactone saponins, namely solanolactosides A, B and two new spirostanol glycosides, namely torvosides M, N. Their structures were characterized as solanolide6-O-[α-L-rhamnopyranosyl-(1 -> 3)-O-β-D-quinovopyranoside], solanolide 6-O-[β-D-xylopyranosyl-(1 -> 3)-O-β-D-quinovopyranoside], yamogenin 3-O-[β-D-glucopyranosyl-(1 -> 6)-O-β-D-glucopyranoside] and neochlorogenin 3-O-[β-D-glucopyranosyl-(1 -> 6)-O-β-D-glucopyranoside] on the basis of spectroscopic analyses.

III. Compounds Isolated from Leaves

Mahmood *et al.,* 1983 isolatedtorvanol A from the leaves. Yuan-Yuan *et al.,* 2011 were able to isolate nine new compounds namely neochlorogenin 6-*O*-*β*-D-quinovopyranoside, neochlorogenin 6-*O*-*β*-D- xylopyranosyl- (1→3) -*β*- D- quinovopyranoside, neochlorogenin 6-*O*-*α*-L-rhamnopyranosyl-(1→3)-*β*-D-quinovopyranoside, solagenin 6-*O*-*β*-D- quinovopyranoside, solagenin 6-*O*-*α*-L- rhamnopyranosyl- (1→3) -*β*-D-quinovopyranoside, isoquercetin, rutin, kaempferol and quercetin. Furostanol glycoside 26-O-beta-glucosidase is an important part of methanolic extracts of the leaves. Mahmood *et al.,* 1983 isolated some non-alkaloidal compounds namely, 3, 4-trimethyltriacontane, octacosanyltriacontanoate and 5-hexatriacontanone by spectral data and chemical studies. Triacontanol, 3-tritriacontanone, tetratriacontanoic acid, sitosterol, stigmasterol and campesterol have also been isolated and identified.

IV. Compounds Isolated from Roots

The root of *Solanum torvum* is a source of steroidal glycosides such astorvosides A-G. Thus, they were structurally characterized as *(25S)-26-O-fl-*

~-glucopyranosyl)-5c~-furostan- 3fl, 6a, 22sc, 26-tetraol 6-O-[a-L-rhamnopyranosyl-(1 ---> 3)-fl-D-quinovopyuranoside], *(25S)-26-O-(fl-D-glucopyranosyl)-22ce-methoxy-Sa-furostan-3fl,6ce,26-triol* 6- O- [fl-D-xylopyranosyl- (1 ---> 3) -/3- D-quinovopyranoside], neosolaspigenin 6-O-[a-L-rhamnopyranosyl-(1 --> 3)-fl-D-quinovopyranoside], neosolaspigenin 6-O-[fl-Dxylopyranosyl-(1 --> 3)-/3-D-quinovopyranoside], *(25S)-26-O-(fl-D-glucopyranosyl)-6a,26-dihydroxy-22a-methoxy*-Sa-furostan-3-one 6-O-(fl-D-quinovopyranoside), *(25S)-26-O-(fl-D-glucopyranosyl)-6a,26-dihydroxy-22a*methoxy- Sce-furostan-3-one 6-O-[/3-D-xylopyranosyl-(1-->3)-/3-D-quinovopyranoside], and (25S)-26-hydroxy-22a-methoxy-5a-furostan-3-one 26-O-(fl-D-glucopyranoside).

Some important phytochemicals isolated from Solanum torvum are:

- Methyl caffeate
- Isoquercetin
- Rutin
- Kaempferol
- Quercetin

METHYL CAFFEATE

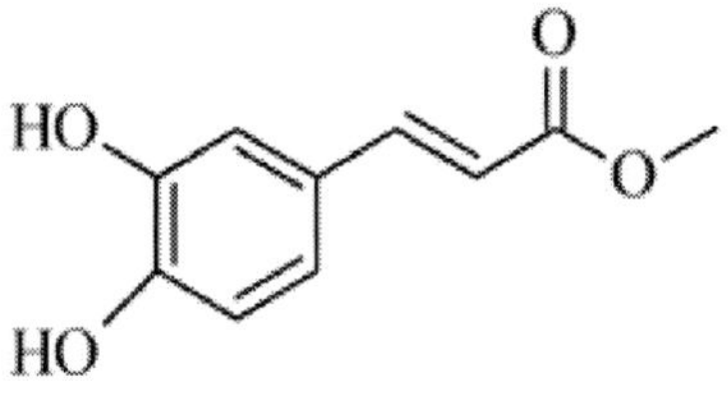

Figure 1. Chemical structure of Methyl caffeate.

IUPAC name:

- Methyl (E)-3-(3,4-dihydroxyphenyl)prop-2-enoate

Other names:

- Caffeic acid methyl ester
- Methylcaffeate
- Methyl 3,4-dihydroxycinnamate

Natural occurrences:

Methyl caffeate is a hydroxycinnamic acid, a naturally occurring phenolic compound. It can be found in *Polygonum amplexicaule var. sinense* (Chinese name: Xuesanqi), a widely distributed medicinal plant in China and in the fruit of Solanum torvum.

Health effect: Methyl caffeate shows an antidiabetic effect in streptozotocin induced diabetic rats.

ISOQUERCETIN

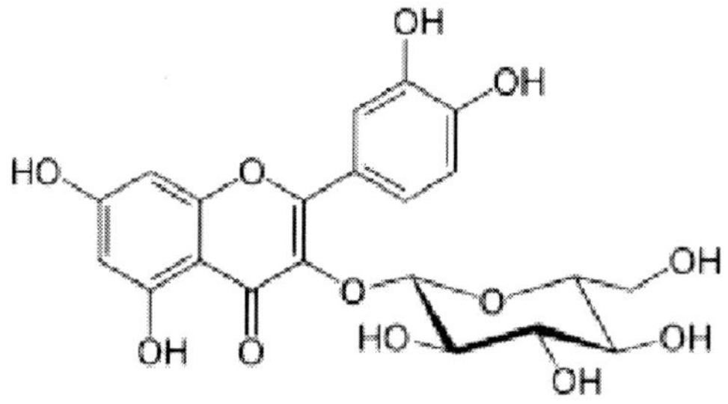

Figure 2. Chemical structure of Isoquercetin.

IUPAC name:

- 2-(3,4-dihydroxyphenyl)-5,7-dihydroxy-3-[(2*S*,3*R*,4*S*,5*S*,6*R*)-3,4,5-trihydroxy-6-(hydroxymethyl)oxan-2-yl]oxychromen-4-one

Other names:

- Isoquercitrin
- Isoquercitroside
- Isoquercitin
- Trifoliin
- Isotrifolin
- trifoliin
- Isohyperoside
- Isotrifoliin
- Quercetin-3-glucoside
- Quercetin-3-O-glucoside
- Quercetin 3-O-β-D-glucopyranoside

Properties:

Molecular formula	$C_{21}H_{20}O_{12}$
Molecular mass	464.37 g/mol

Isoquercitin is a chemical compound. It can be isolated from mangoes and *Rheum nobile*, the Noble rhubarb or Sikkim rhubarb, a giant herbaceous plant native to the Himalaya. Quercetin glycosides are also present in tea.

RUTIN

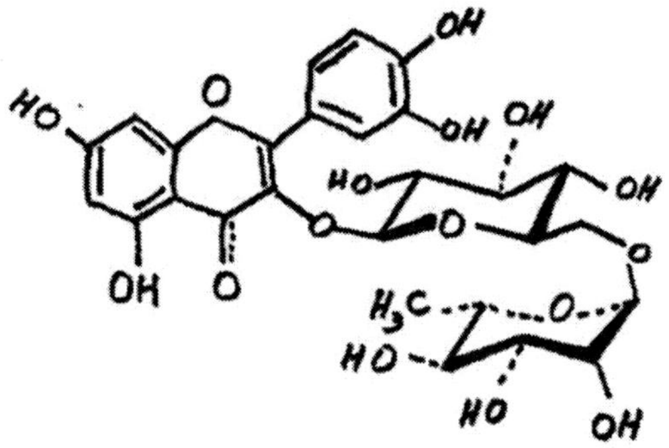

Figure 3. Chemical structure of Rutin.

IUPAC name:

- 2-(3,4-dihydroxyphenyl)-5,7-dihydroxy-3-{[(2*S*,3*R*,4*S*,5*S*,6*R*)-3,4,5-trihydroxy-6-({[(2*R*,3*R*,4*R*,5*R*,6*S*)-3,4,5-trihydroxy-6-methyloxan-2-yl]oxy}methyl)oxan-2-yl]oxy}-4*H*-chromen-4-one

Preferred IUPAC name:

- 2-(3,4-dihydroxyphenyl)-5,7-dihydroxy-3-[α-L-rhamnopyranosyl-(1→6)-β-D-glucopyranosyloxy]-4*H*-chromen-4-one

Other names:

- Rutoside
- Phytomelin
- Sophorin
- Birutan
- Eldrin

- Birutan forte
- Rutin trihydrate
- Globularicitrin

Properties:

Molecular formula	$C_{27}H_{30}O_{16}$
Molecular mass	610.517 g/mol
Appearance	Solid
Melting point	242 °C, 515 K, 468 °F

Rutin, also known as rutoside, quercetin-3-O-rutinoside and sophorin, is the glycoside between the flavonolquercetin and the disacchariderutinose (α-L-rhamnopyranosyl-(1→6)-β-D-glucopyranose).

Occurrences:

Rutin is a citrusflavonoidglycoside found in buckwheat, the leaves and petioles of *Rheum* species, as well as asparagus. Rutin is also found in the fruits and flowers of the pagoda tree, fruits and fruit rinds [especially the citrus fruits (orange, grapefruit, lemon, lime)] and berries such as mulberry, ash tree fruits and cranberries. Its name come from *Ruta graveolens*, a plant that also contains rutin. It is sometimes referred to as vitamin P, although not strictly a vitamin. Rutin is one of the primary flavonols found in 'clingstone'peaches.

Chemical relatives:

Rutin (quercetin rutinoside), as well as quercitrin, is a glycoside of the flavonoid quercetin. As such, the chemical structures of both are very similar, with difference existing in the hydroxyl functional group. Both quercetin and rutin are used in many countries as medications for blood vessel protection, and are ingredients of numerous multivitamin preparations and herbal remedies. Rutin is found in many plants, especially the buckwheat plant *Fagopyrum esculentum* Moench, family of Polygonaceae.

Health effects:

While a body of evidence for the effects of rutin and quercetin are available in mice, rats' hamsters and rabbits, as well as *in vitro* studies, no clinical studies directly demonstrate significant, positive effects of rutin as dietary supplement in humans.

- Rutin inhibits platelet aggregation, as well as decreases capillary permeability, making the blood thinner and improving circulation.

- Rutin shows anti-inflammatory activity in some animal and *in vitro* models.
- Rutin inhibits aldose reductase activity. Aldose reductase is an enzyme present in the eye and elsewhere in the body. It helps to convert glucose into sugar alcohol sorbitol.
- Rutin also strengthens the capillaries, and, therefore, can reduce the symptoms of haemophilia. It also may help to prevent a common, unpleasant-looking, venous edema of the legs. However, a double-blind clinical study on the effect of buck wheat on tea containing rutin did not show a significant effect above placebo.
- Rutin, as ferulic acid, can reduce the cytotoxicity of oxidized LDL cholesterol and lower the risk of heart disease. Recent studies show that rutin could help prevent blood clots, therefore, could be used to treat patients at risk of heart attacks and strokes
- Some evidence also shows that rutin can be used to treat hemorrhoids, varicosis, and microangiopathy.
- Rutin is also an antioxidant; compared to quercetin, acacetin, morin, hispidulin, hesperidin, and naringin, thus it was found to be the strongest. However, in other trials, the effects of rutin were lower or negligible compared to those of quercetin.

Hydroxyethylrutosides, synthetic hydroxyethyl acetylations of rutin, were used in the treatment of chronic venous insufficiency.

In veterinary medicine:

Rutin has a veterinary use in the management of chylothorax in dogs and cats.

Metabolism:

The enzyme quercitrinase can be found in *Aspergillus flavus*. It is an enzyme in the rutin catabolic pathway.

KAEMPFEROL

IUPAC name:

- 3,5,7-Trihydroxy-2-(4-hydroxyphenyl)-4*H*-chromen-4-one

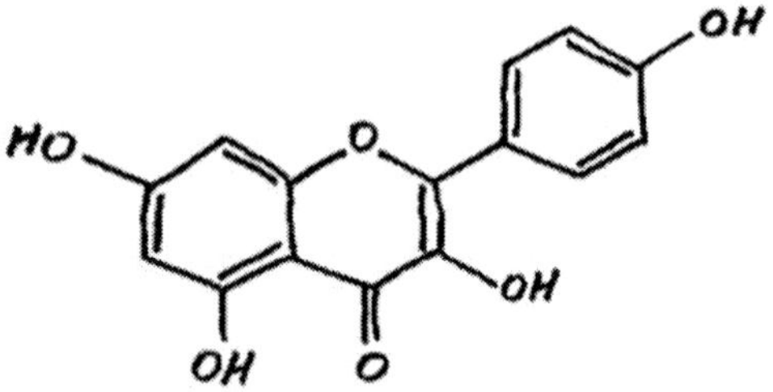

Figure 4. Chemical structure of Kaempferol.

Other names:

- Kaempherol
- Robigenin
- Pelargidenolon
- Rhamnolutein
- Rhamnolutin
- Populnetin
- Trifolitin
- Kempferol
- Swartziol
- 3,5,7-trihydroxy-2-(4-hydroxyphenyl)-4*H*-1-benzopyran-4-one

Properties:

Molecular formula	$C_{15}H_{10}O_6$
Molecular mass	286.23 g/mol

Chemistry:

Kaempferol is a natural flavonol, a type of flavonoid. Kaempferol is a yellow crystalline solid with a melting point of 276 – 278°C. It is slightly soluble in water but soluble in hot ethanol and diethyl ether.

Natural occurrences:

It can be isolated from tea, broccoli, *Delphinium*, Witch-hazel, grapefruit, cabbage, kale, beans, endive, leek, tomato, strawberries, grapes, brussels sprouts, apples and other plant sources.

Kaempferol and its glucoside can be isolated from the methanolic extract of fronds of the fern *Phegopteris connectilis*.

Biological activity:

Some epidemiological studies have found a positive association between the consumption of foods containing kaempferol and a reduced risk of developing several disorders such as cancer and cardiovascular diseases. Numerous preclinical studies have shown that kaempferol and some glycosides of kaempferol have a wide range of pharmacological activities, suich as antioxidant, anti-inflammatory, antimicrobial, anticancer, cardio-protective, neuroprotective, antidiabetic, anti-osteoporotic, estrogenic /antiestrogenic, anxiolytic, analgesic and antiallergic activities).

- Many glycosides of kaempferol, such as kaempferitrin and astragalin, have been isolated as natural products from plants. Kaempferol consumption in tea and broccoli has been associated with reduced risk of heart disease. A related compound, Naringenin may be present at the same time, such as in grapefruit.
- Kaempferol gives the flowers of *Acacia decurrens* and *Acacia longifolia* their color. Antidepressant properties have been reported in tests on animals.
- An 8-year study found that the three flavonols (kaempferol, quercetin, and myricetin) reduced the risk of pancreatic cancer by 23 percent.
- Kaempferol consumption appears to reduce lung cancer incidence.
- Kaempferol may be a potent prophylactic against NOX-mediated neurodegeneration.

Metabolism:

The enzyme kaempferol 4'-O-methyltransferase uses S-adenosyl methionine and kaempferol to produce S-adenosylhomocysteine and kaempferide.

Glycosides:

- Astragalin (Astragaline, asragalin, kaempferol-3-O-glucoside)
- Kaempferitrin (Kaempferol 3,7-dirhamnoside)
- Kaempferol-3-O-β-D-glucopyranoside-7-O-α-L-rhamnopyranoside, a known bitter-tasting flavonoid glycoside, can be isolated from the rhizomes of *Selliguea feei*.
- Kaempferol 7-O-glucoside
- Robinin (Kaempferol-3-O-robinoside-7-O-rhamnoside)
- Sophoraflavonoloside (kaempferol 3-O-sophoroside)
- Trifolin (Kaempferol-3-O-galactoside)

Quercetin

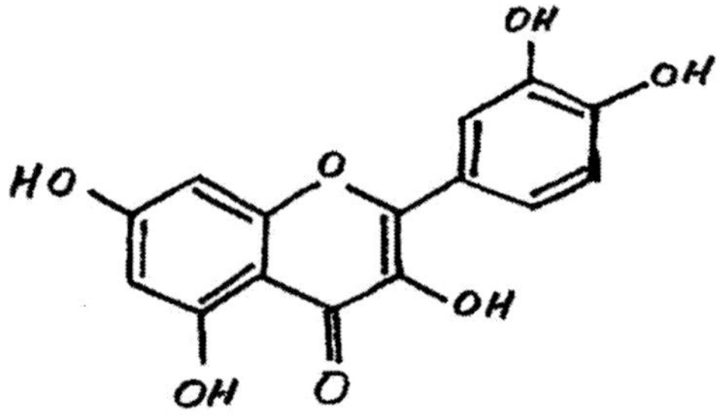

Figure 5. Chemical structure of Quercetin.

IUPAC name:

- 2-(3,4-dihydroxyphenyl)-3,5,7-trihydroxy-4*H*-chromen-4-one

Other names:

- Sophoretin
- Meletin
- Quercetine
- Xanthaurine
- Quercetol
- Quercitin
- Quertine

Properties:

Molecular formula	$C_{15}H_{10}O_7$
Molar mass	302.236 g/mol
Appearance	yellow crystalline powder
Density	1.799 g/cm3
Melting point	316 °C

Occurrence:

Quercetin is a flavonoid widely distributed in nature. The name has been in use since 1857, and is derived from *quercetum* (oak forest), after *Quercus*. It is a naturally-occurring polar auxin transport inhibitor.

Foods rich in quercetin include black and green tea (*Camellia sinensis*; 2000–2500 mg/kg), capers (1800 mg/kg), lovage (1700 mg/kg), apples

(440 mg/kg), onion, especially red onion (191 mg/kg) (higher concentrations of quercetin occur in the outermost rings), red grapes, citrus fruit, tomato, broccoli and other leafy green vegetables, and a number of berries, such as raspberry, bog whortleberry (158 mg/kg, fresh weight), lingonberry (cultivated 74 mg/kg, wild 146 mg/kg), cranberry (cultivated 83 mg/kg, wild 121 mg/kg), chokeberry (89 mg/kg), sweet rowan (85 mg/kg), rowanberry (63 mg/kg), sea buckthorn berry (62 mg/kg), crowberry (cultivated 53 mg/kg, wild 56 mg/kg), and the fruit of the prickly pear cactus. A recent study found that organically grown tomatoes had 79% more quercetin than "conventionally grown". A study by the University of Queensland, Australia has also indicated the presence of quercetin in varieties of honey, such as honey derived from eucalyptus and tea treeflowers.

Biosynthesis:

Phenylalanine is converted to 4-coumaroyl-CoA in a series of steps known as the general phenylpropanoid pathway using phenylalanine ammonia-lyase, cinnamate – 4 - hydroxylase, and 4 – coumaroyl - CoA-ligase. 4 - Coumaroyl-CoA is added to three molecules of malonyl-CoA to form tetrahydroxychalcone using 7, 2 – dihydroxy – 4 - methoxyisoflavanol synthase. Tetrahydroxychalcone is then converted into naringenin using chalcone isomerase. Naringenin is then converted into eriodictyol using flavanoid 3 - hydroxylase. Eriodictyol is also converted into dihydroquercetin with flavanone 3-hydroxylase, which is then converted into quercetin using flavonol synthase. Quercetin is release in the rutin degradation pathway. The enzyme quercitrinase can be found in Aspergillus flavus. Its substrates are quercitrin and H2O and releases quercetin and L-rhamnose. It is an enzyme in the rutin catabolic pathway.

Conjugates:

Quercetin 3-O-sulfate is a human plasma quercetin metabolite.

Glycosides:

Quercetin is the aglycone form of a number of other flavonoid of glycosides, such as rutin and quercitrin, found in citrus fruit, buckwheat and onions. Quercetin forms the glycosidesquercitrin and rutin together with rhamnose and rutinose, respectively. Likewise guaijaverin is 3-O-arabinoside, hyperoside is 3-O-galactoside, isoquercitin is 3-O-glucoside and spiraeoside is 4'-O-glucoside. CTN-

986 is a quercetin derivative found in cotton seeds and cotton seed oil. Miquelianin is the quercetin 3-o-β-d-glucuronopyranoside.

Study of metabolism *using Cunninghamella elegans:*

Cunninghamella elegans is a fungal model of mammalian drug metabolism. It can be used to study the quercetin biodegradation.

Effects of consumption by humans and other animals:

Quercetin itself (aglycone quercetin), as opposed to quercetin glycosides, is not a normal dietary component. In a bioavailability study in rats, radiolabelled quercetin-4'-glucoside was converted to phenolic acids as it passed through the gastrointestinal tract, producing compounds not monitored in several previous animal studies of aglycone quercetin. All but 4% of the radiolabel was recovered within 72 hours, with 69% recovered in urine. Quercetin has neither been confirmed scientifically as a specific therapeutic for any condition nor been approved by any regulatory agency. The U.S. Food and Drug Administration have not approved any health claims for quercetin.

Preliminary research:

Antiviral

In 2007, a study assessed the anti-Hepatitis B effects of Hyperoside, and was published in the Acta Pharmacologica Sinica, it was shown that Hyperoside (which is the 3-O-galactoside of quercetin) is a strong inhibitor of HBsAg and HBeAg secretion in 2.2.15 cells.

In another study also published in 2007 in the Archives of Pharmacal Research it was shown that quercetin, quercitrin and myricetin 3-O-beta-D-galactopyranoside displayed inhibition against HIV-1 reverse transcriptase, all with IC50 values of 60 microM.

Cancer

The American Cancer Society says while quercetin "has been promoted as being effective against a wide variety of diseases, such as cancer,some early lab results appear promising, and yet there is no reliable clinical evidence that quercetin can prevent or treat cancer in humans." In the amounts consumed in a healthy diet, quercetin "is unlikely to cause any major problems or benefits". In laboratory studies of cells in vitro, quercetin produces changes that are also produced by compounds that cause cancer (carcinogens), nevertheless, these studies do not report increased cancer in animals or humans.

From laboratory studies, it was conjectured that quercetin may affect certain mechanisms of cancer. However, a research was carried out for a period of 8 years and found that the presence of three flavonolskaempferol, quercetin, and myricetin in a normal diet was associated with 23% reduced risk of pancreatic cancer, a rare but frequently fatal disease, in tobacco smokers. There was no benefit in subjects who had never smoked or had previously quit smoking.

In vitro, cultured skin and prostate cancer cells were suppressed (compared to nonmalignant cells) when treated with a combination of quercetin and ultrasound. In laboratory cultural studies, quercetin increased the sensitivity of resistant colorectal tumors with microsatellite instability to the chemotherapy drug, 5-fluorouracil.

Inflammation

Several laboratory studies have shown that quercetin may have anti-inflammatory properties, and it is being investigated for a wide range of potential health benefits. Quercetin has been reported to be of importance in alleviating symptoms of pollinosis. An enzymatically modified derivative was found to alleviate ocular but not nasal symptoms of pollinosis. Studies done in test tubes have shown that quercetin may affect immune cells from releasing histamines which might influence symptoms of allergies.A study with rats showed that quercetin effectively reduced immediate-release of niacin (vitamin B_3) flush, in part by means of reducing prostaglandin D2 production. A pilot clinical study of four humans gave preliminary data supporting this. Quercetin may have properties of a calcineurin inhibitor, similar to cyclosporin A and tacrolimus, according to one laboratory study.

Fibromyalgia

Quercetin is said to be effective in the treatment of fibromyalgia because of its potential anti-inflammatory or mast cell inhibitory properties shown in laboratory studies.

Metabolic syndrome

Quercetin has been shown to increase energy expenditure in rats, only for short periods (fewer than 8 weeks). Effects of quercetin on exercise tolerance in mice are associated with increased mitochondrial biogenesis. In mice, an oral quercetin dose of 12.5 to 25 mg/kg increased gene expression of mitochondrialbiomarkers and improved exercise endurance. It has also been claimed that quercetin reduces blood pressure in hypertensive and obese subjects in whom LDL cholesterol levels were also reduced. An in vitro study showed that quercetin and resveratrol combine to inhabit the production of fat cells.

A 12-week study of 941 adults found that supplements of 500 to 1000 milligrams of quercetin with vitamin C and niacin did not cause any significant difference in body mass or composition and had no significant effect on inflammatory markers, diagnostic blood chemistries, blood pressure, and blood lipid profiles.

Drug interactions

Quercetin is contraindicated with some antibiotics; it may interact with fluoroquinolones (an antibiotic), as quercetin competitively binds with bacterial DNA gyrase. Although it is not certain whether it inhibits the effect of fluoroquinolones.

CAFFEIC ACID

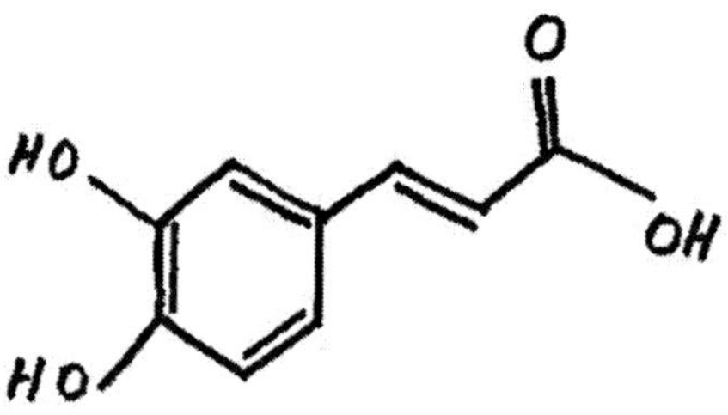

Figure 6. Chemical structure of Caffeic acid.

IUPAC name:

- 3-(3,4-Dihydroxyphenyl 2-propenoic acid
- 3,4-Dihydroxy-cinnamic acid

- trans-Caffeate
- 3,4-Dihydroxy-trans-cinnamate)
- (E)-3-(3,4-dihydroxyphenyl)-2-propenoic acid
- 3,4-Dihydroxybenzeneacrylicacid
- 3-(3,4-Dihydroxyphenyl)-2-propenoic acid

Properties:

Molecular formula	$C_9H_8O_4$
Molar mass	180.16 g/mol
Density	1.478 g/cm^3
Melting point	223–225 °C

Caffeic acid is an organic compound that has been classified as a hydroxycinnamic acid. This yellow solid consists of both phenolic and acrylic functional groups, is found in all plants because it is a key intermediate in the biosynthesis of lignin, one of the principal sources of biomass.

Occurrence:

Caffeic acid can be found in the bark of *Eucalyptus* globulus, as well as in the freshwater fern *Salvinia molesta* or in the mushroom *Phellinus linteus.* Caffeic acid is one of the main natural phenols in argan oil.

Biosynthesis:

Caffeic acid, which is unrelated to caffeine, is biosynthesized by hydroxylation of coumaroyl ester of quinic ester. This hydroxylation produces the caffeic acid ester of shikimic acid, which converts to chlorogenic acid. It is the precursor to ferulic acid, coniferyl alcohol, and sinapyl alcohol, all of which are significant building blocks in lignin. Its transformation into ferulic acid is catalyzed by the enzyme caffeic acid-O-methyltransferase. Caffeic acid and its derivative caffeic acid phenethyl ester (CAPE) are produced in many kinds of plants.

Biotransformation:

Caffeate O-methyltransferase is an enzyme responsible for the transformation of caffeic acid into ferulic acid.

Biodegradation:

Caffeate 3,4-dioxygenase is an enzyme that uses caffeic acid and oxygen to produce 3-(2-carboxyethenyl)-cis,cis-muconate

Glycosides:

3-O-caffeoylshikimic acid (dactylifric acid) and its isomers, are enzymic browning substrates found in dates (*Phoenix dactylifera* fruits).

Pharmocology:

Caffeic acid has a variety of potential pharmacological effects in *in vitro* studies and in animal models, but no clinical effects have been identified in humans. Caffeic acid is an antioxidant *in vitro* and also *in vivo.* It also shows immunomodulatory and antiinflammtory activity. Similarly, it outperformed the other antioxidants, reducing aflatoxin production by more than 95 percent. The studies were the first to show that oxidative stress that would otherwise trigger or enhance *Aspergillus flavus* aflatoxin production can be stymied by caffeic acid. This opens the door to using natural fungicide methods by supplementing trees with antioxidants.

Other uses

Caffeic acid is the active ingredient in caffenol, a do-it-yourself black-and-white photographic developer made from instant coffee. This developing chemistry is similar to that of catechol or pyrogallol. It is also used as a matrix in MALDI mass spectrometry analyses.

GALLIC ACID

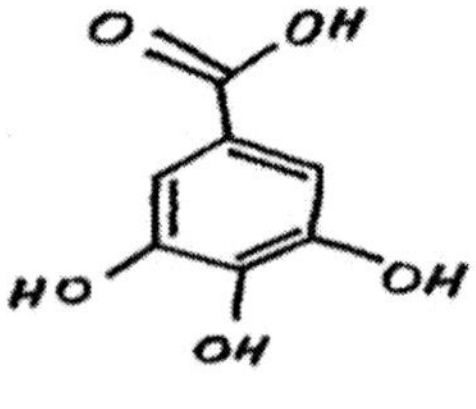

Figure 7. Chemical structure of Gallic acid;

IUPAC name:

- 3,4,5-trihydroxybenzoic acid

Other names:

- Gallic acid
- Gallate
- 3,4,5-trihydroxybenzoate

Properties:

Molecular formula	$C_7H_6O_5$
Molar mass	170.12 g/mol
Appearance	White, yellowish-white, or pale fawn-colored crystals.
Density	1.7 g/cm3 (anhydrous)
Melting point	250 °C, 523 K, 482 °F

Gallic acid is a trihydroxybenzoic acid, a type of phenolic acid, a type of organic acid, also known as 3,4,5-trihydroxybenzoic acid, found ingallnuts, sumac, witch hazel, tea leaves, oak bark, and other plants. The chemical formula is $C_6H_2(OH)_3COOH$. Gallic acid is found to be free and also as part of tannins. Salts and esters of gallic acid are termed 'gallates'. Despite, it does not contain gallium.

Gallic acid is commonly used in the pharmaceutical industries. It is used as a standard for determining the phenol content of various analytes by the Folin-Ciocalteau assay, results are reported in gallic acid equivalents. Gallic acid can also be used as a starting material in the synthesis of the psychedelic alkaloid mescaline.

Thus, Gallic acid seems to have anti-fungal and anti-viral properties. Hence, it acts as an antioxidant and helps to protect human cells against oxidative damage. Gallic acid was found to show cytotoxicity against cancer cells, without harming healthy cells. As a result, it is used as a remote astringent in cases of internal haemorrhage. In the same vein, it is also used to treat albuminuria and diabetes. Some ointments to treat psoriasis and external haemorrhoids contain gallic acid.

Natural occurrence:

Gallic acid is found in a number of land plants. It is also found in the aquatic plant Myriophyllum spicatum and shows an allelopathic effect on the growth of the blue-green alga Microcystis aeruginosa.

Metabolism:

Gallate 1-beta-glucosyltransferase (EC 2.4.1.136) is an enzyme that uses UDP-glucose and gallate, whereas its two products are UDP and 1-galloyl-beta-D-glucose. Gallate decarboxylase (EC 4.1.1.59) is another enzyme.

Health effects:

It is a weak carbonic anhydrase inhibitor.

PYROGALLOL

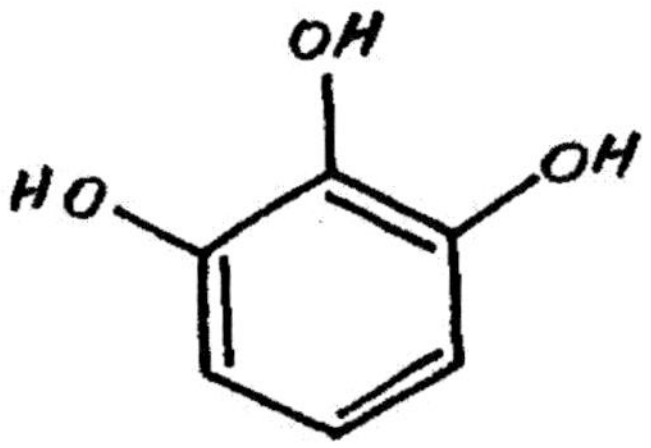

Figure 8. Chemical structure of Pyrogallol.

Other names:

- 1,2,3-Trihydroxybenzene
- Pyrogallic acid

Properties:

Molecular formula	$C_6H_6O_3$
Molar mass	126.11 g/mol
Density	1.45 g/cm3
Melting point	131–134 °C
Boiling point	309 °C
Natural occurrences:	

Pyrogallol or benzene-1,2,3-triol is a benzenetriol. The aquatic plant Myriophyllum spicatum produces pyrogallic acid.

Chemistry:

It is a white crystalline powder and a powerful reduction agent. It was first prepared by Scheele in 1786 by heating gallic acid. An alternative preparation is by heating para-chlorophenoldisulphonic acid with potassium hydroxide. When in alkaline solution, it absorbs oxygen from the air, turning purple from a colourless solution. It can also be used in this way to calculate the amount of oxygen in air, notably via the use of the Orsat apparatus.

Uses:

One can find its uses in hair dying, dying of suturing materials and oxygen absorption in gas analysis. It also has antiseptic properties. Pyrogallol was also used as a developing agent in black-and-white developers, but its use is largely historical except for special purpose applications (Hydroquinone is more commonly used today). Harmful if swallowed, and repeated or prolonged exposure to this compound is not recommended. Pyrogallol is not likely to be used as a modern hair dye due to the suspected toxicity of the compound.

ARACHIDIC ACID

Arachidic acid, also called eicosanoic acid is the saturated fatty acid with a 20 carbon chain. It serves as a minor constituent of peanut oil (1.1%–1.7) and corn oil (3%). Its name was derived from the Latin *arachis*peanut. It can be formed by the hydrogenation of arachidonic acid. Reduction of arachidic acid yields arachidyl alcohol. Arachidic acid is used for the production of detergents, photographic materials and lubricants.

Figure 9. Chemical structure of Arachidic acid.

IUPAC Name:

- Icosanoic

Other names:

- eicosanoic acid
- *n*-eicosanoic acid
- arachidic acid
- arachic acid

***Properties*:**

Molecular formula	$C_{20}H_{40}O_2$
Molecular mass	312.53 g mol^{-1}
Appearance	White crystalline solid
Density	0.8240 g/cm^3
Melting point	75.5 °C, 349 K, 168 °F
Boiling point	328 °C,601 K,622 °F

BETA-SITOSTEROL

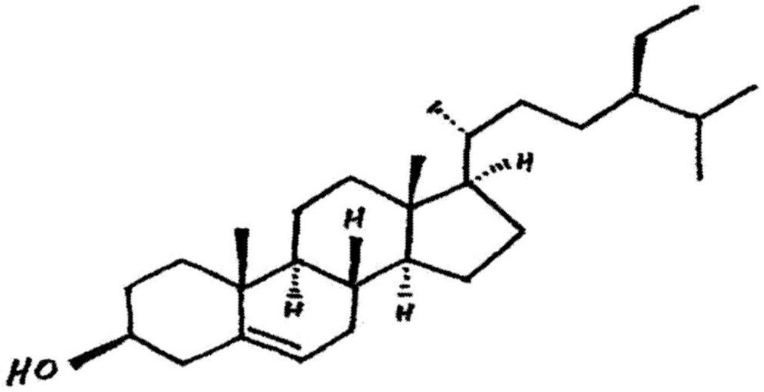

Figure 10. Chemical structure of *beta*-Sitosterol.

IUPAC name:

- 17-(5-Ethyl-6-methylheptan-2-yl)-10
- 13-dimethyl-2,3,4,7,8,9,11,12,14,15,16,17-dodecahydro-1*H*-cyclopenta[*a*]phenanthren-3-ol

Other names:

- 22,23-Dihydrostigmasterol
- Stigmast-5-en-3-ol
- β-Sitosterin

Properties:

Molecular formula	$C_{29}H_{50}O$
Molecular mass	414.71 g mol^{-1}
Melting point	136–140 °C

β-Sitosterol is one of several phytosterols (plant sterols) with chemical structures similar to that of cholesterol. Sitosterols are white, waxy powders with a characteristic odor. They are hydrophobic and soluble in alcohols.

Sources:

It is widely distributed in the plant kingdom and found in *Nigella sativa*, pecans, *Serenoa repens* (saw palmetto), avocados, *Cucurbita pepo* (pumpkin seed), *Pygeum africanum*, cashew fruit, rice bran, wheat germ, corn oils, soybeans, sea-buckthorn, wolfberries, and *Wrightia tinctoria*.

Medicinal Uses:

It could be used alone or in combination with similar phytosterols. β-sitosterol reduces blood levels of cholesterol, and is sometimes used in treating hypercholesterolemia. β-Sitosterol inhibits cholesterol absorption in the intestine. When the sterol is absorbed in the intestine, it is transported by lipoproteins and incorporated into the cellular membrane. Phytosterols and phytostanols both inhibit the uptake of dietary and biliary cholesterol, thereby decreasing the levels of LDL and serum total cholesterol. Because the structure of β-sitosterol is very similar to that of cholesterol, β-sitosterol takes the place of dietary and biliary cholesterol in micelles produced in the intestinal lumen. This causes less cholesterol absorption in the body.

A similar study in this regard has shown a positive effect on male hair loss in combination with Saw palmetto. In Europe, β-sitosterol is used in herbal therapy, especially in the treatment of benign prostatic hyperplasia (BPH). In-vitro studies on a cancer cell line suggest potential for the treatment of prostatic carcinoma.

Chemical engineering:

The use of sitosterol as a chemical intermediate was for many years limited due to the lack of a chemical point of attack on the side-chain that would permit its removal. Extensive efforts on the part of many laboratories eventually led to the discovery of a pseudomonasmicrobe that efficiently affected that transformation. Fermentation digests the entire aliphatic side-chain at carbon 17 to afford a mixture of 17-keto products including dehydroepiandrosterone.

Side-Effects:

- β-Sitosterol should be avoided during pregnancy and breast-feeding, since not enough is known about its effects on unborn and newborn children.
- β-Sitosterol is also not recommended for individuals with sitosterolemia, a rare inherited fat storage disease. Because people with this condition have too much β-sitosterol and related fats in their system, taking β-sitosterol will only worsen this condition.
- High levels of β-sitosterol concentrations in the blood have been correlated with increased severity of heart disease in men who previously have suffered from heart attacks.

Genetic disorder:

While plant sterols are usually beneficial, there is a rare autosomal recessive genetic disorder phytosterolemia which causes over-absorption of phytosterols. Phytosterols accumulation and over-absorption in tissue, causes premature coronary artery disease and tendon xanthoma.

Precursor of anabolic steroid Boldenone:

Being a steroid, β-sitosterol is a precursor of anabolic steroid Boldenone. Boldenone undecylenate is commonly used in veterinary medicine to induce growth in cattle. However, it is also one of the most commonly abused anabolic steroids in sports. This leads to suspecting that some athletes who test positive on Boldenone undecylenate didn't actually abuse the hormone itself but consumed food rich in β-sitosterol.

Synthesis:

Total synthesis of β-sitosterol has not been achieved. However, β-sitosterol has been synthesized from the following. Stigmasterol, which involves a specific hydrogenation of the side-chain of stigmasterol. The first step in the synthesis forms stigmasterol tosylate.

From stigmasterol **1** (95% purity) using p-TsCl, DMAP, and pyridine (90% yield). The tosylate then undergoes solvolysis as it is treated with pyridine and anhydrous MeOH to give a 5:1 ratio of i-stigmasterol methyl ether **3** (74% yield) to stigmasterol methyl ether **4**, which is subsequently removed by chromatography. The hydrogenation step of a previously proposed synthesis involved the catalyst Pd/C and the solvent ethyl acetate. However, due to isomerisation during hydrolysis, other catalysts, such as PtO2, and solvents, such as ethanol, were tested. There was little change with the use of a different catalyst. Ethanol, however, prevented isomerisation and the formation of the unidentified impurity to give compound **5**. The last step of the synthesis is unprotection of the β-ring double bond of **5** with p-TsOH, aqueous dioxane, and heat (80 °C) to yield β-sitosterol **6**. The cumulative yield for the finaltwo steps was 55%, and the total yield for the synthesis was 37%.

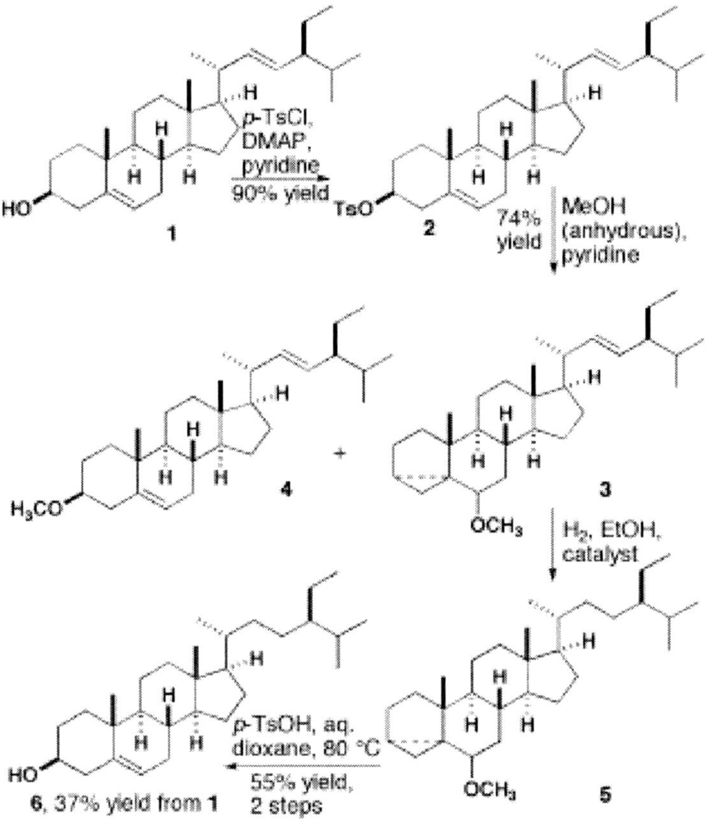

Biosynthesis:

The regulation of the biosynthesis of both sterols and some specific lipids occurs during membrane biogenesis. Through 13C-labeling patterns, it has been determined that both the mevalonate and deoxyxylulose pathways were involved in the formation of β-sitosterol. The precise mechanism of β-sitosterol formation varies according to the organism, but is generally found to come from cycloartenol. The biosynthesis of cycloartenol begins as one molecule of isopentenyl diphosphate (IPP) and two molecules of dimethylallyl diphosphate (DMAPP) form farnesyl diphosphate (FPP). Two molecules of FPP were then joined tail-to-tail to yield squalene, a triterpene. Squalene, through a cyclization reaction with 2,3-oxidosqualene 6 as an intermediate forms of cycloartenol.

The biosynthesis of β-sitosterol from cycloartenol is summarized below.

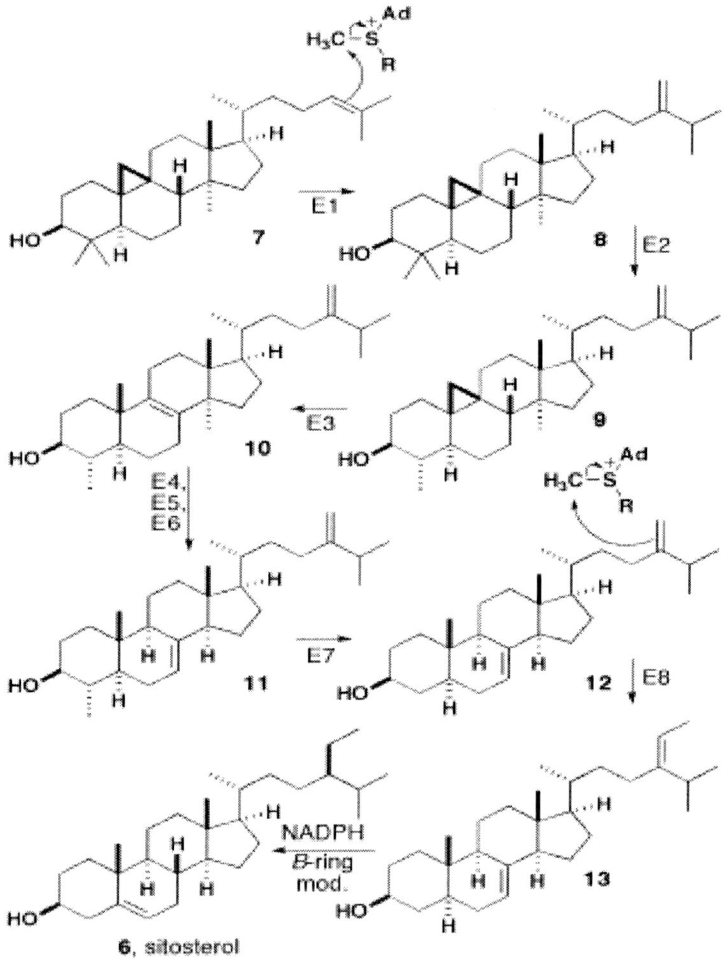

Figure 10. Biosynthesis of β-sitosterol (6) from cycloartenol (7).

The double bond of cycloartenol (compound 7 in diagram) is methylated by SAM to give a carbocation that undergoes a hydride shift and loses a proton to yield a compound with a methylene side-chain. Both of these steps were catalyzed by sterol C-24 methyltransferase (Step E1 in diagram). Compound 8 is then catalyzed by sterol C-4 demethylase (E2) and loses a methyl group to produce cycloeucalenol. Subsequent to this, the cyclopropane ring is opened with cycloeucalenol cycloisomerase (E3) to form **10**. Compound **10** loses a methyl group and undergoes an allylic isomerization to form Gramisterol **11**. This step is catalyzed by sterol C-14 demethylase (E4), sterol Δ14-reductase (E5), and sterol Δ8-Δ7-isomerase (E6). The last methyl group is removed by sterol demethylase (E7) to form episterol **12**. Episterol **12** is methylated by SAM to produce a second carbocation, which loses a proton to yield **13**. This step is catalyzed by 24-methylenesterol C-methyltransferase (E8). Compound **13** then undergoes reduction by NADPH and modifications in the β-ring to form β-sitosterol.

CHACONINE

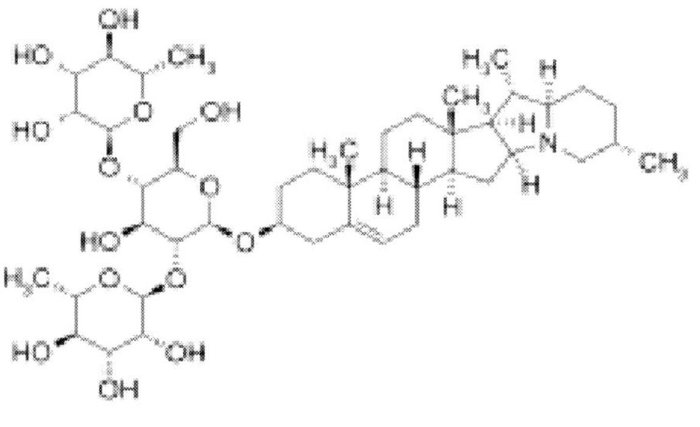

Figure 11. Chemical structure of Chaconine.

Properties:

Molecular formula	$C_{45}H_{73}NO_{14}$
Molecular mass	852.06 g mol^{-1}
Melting point	243°

α-Chaconine is a steroidalglycoalkaloidchemicalcompound that occurs in plants of the Solanaceae family. It is a natural toxicant produced in green potatoes and gives the potato a bitter taste.

MONTANIC ACID

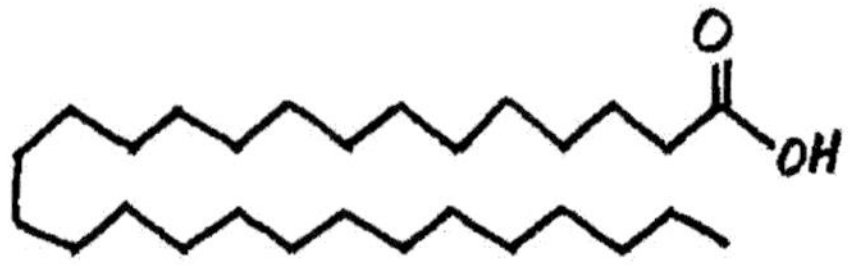

Figure 12. Chemical structure of Montanic acid.

IUPAC Name:

- Octacosanoic acid

Properties:

Molecular formula	$C_{28}H_{56}O_2$
Molecular mass	424.74 g mol^{-1}
Density	0.8191 g/mL
Melting point	90.9 °C, 364 K, 196 °F

Montanic acid is a saturated fatty acid isolated and detected mainly in montan wax. Also occurs in beeswax and Chinese wax. Montanic acid ethylene glycolesters and glycerol esters are used as protective layer on fruit skins and coating on foods.

SOLANINE

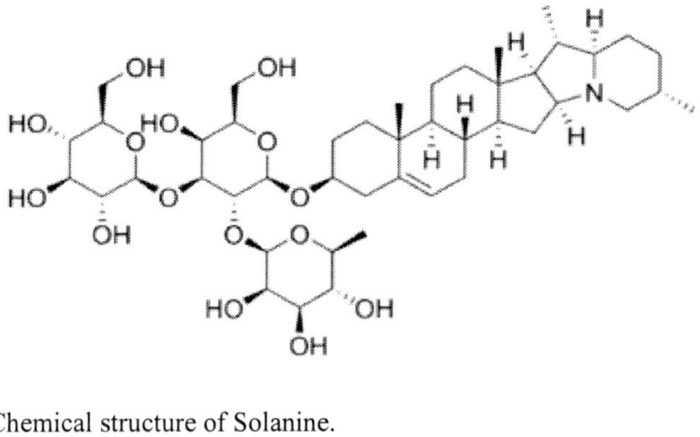

Figure 13. Chemical structure of Solanine.

Properties:

Molecular formula	$C_{45}H_{73}NO_{15}$
Molecualr mass	868.06
Appearance	White crystalline soild
Melting point	271 - 273 °C

Solanine is a glyco alkaloidpoison found in species of the nightshade family (Solanaceae), such as the potato (*Solanum tuberosum*). It can occur naturally in any part of the plant, including the leaves, fruit, and tubers. Solanine has fungicidal and pesticidal properties, and it is one of the plant's natural defenses. Solanine was first isolated in 1820 from the berries of the European black nightshade (*Solanum nigrum*), after which it was named.

Solanine poisoning:

Symptoms:

Solanine poisoning is primarily displayed by gastrointestinal and neurological disorders. Symptoms include nausea, diarrhea, vomiting, stomach cramps, burning of the throat, cardiac dysrhythmia, headache and dizziness. In more severe cases, hallucinations, loss of sensation, paralysis, fever, jaundice, dilated pupils, hypothermia and death have been reported.

Solanine poisoning if taken in large quantity can result to death. One of the studies suggested that doses of 2 to 5 mg per kilogram of body weight can cause toxic symptoms, and doses of 3 to 6 mg per kilogram of body weight could be fatal. Symptoms usually occur within 8 to 12 hours after ingestion, but may occur as rapidly as 30 minutes after eating high-solanine foods. The lowest dose that could cause symptoms of nausea is about 25 mg solanine for adults; a life-threatening dose for a regular-weight adult ranges about 400 mg solanine

Mechanism of action:

One study suggests that the toxic mechanism of solanine is caused by the chemical's interaction with mitochondrial membranes. Experiments showed that solanine exposure opens the potassium channels of mitochondria, decreasing their membrane potential. This in turn leads to Ca^{2+} being transported from the mitochondria into the cytoplasm, and it is this increased concentration of Ca^{2+} in the cytoplasm that triggers cell damage and apoptosis.

Correlation with birth defects:

Some studies have shown a correlation between the consumption of potatoes suffering from late blight (which increases solanine and other glycoalkaloid levels) and the incidence of congenitalspina bifida in humans. However, other studies have shown no correlation between potato consumption and the incidence of birth defects.

Solanine in potatoes:

Solanine occurs naturally in many species of the genus *Solanum*, such as the potato (*Solanum tuberosum*), tomato (*Solanum lycopersicum*), eggplant (*Solanum melongena*), and bittersweet nightshade (*Solanum dulcamara*). Potatoes naturally produce solanine and chaconine, a related glycoalkaloid, as a defense mechanism against insects, disease, and predators. Potato leaves, stems and shoots are naturally high in glycoalkaloids.

When potato tubers are exposed to light, they turn green and increase glycoalkaloid production. This is a natural defense to help prevent the uncovered tuber from being eaten. The green colour shows the presence of chlorophyll, and it is harmless. However, it is an indication that increased level of solanine and chaconine may be present. Some diseases, such as late blight, can dramatically increase the levels of glycoalkaloids present in potatoes. Mechanically damaged potatoes also produce increased levels of glycoalkaloids. This is believed to be a natural reaction of the plant in response to disease and damage. In potato tubers, 30–80% of the solanine develops in and close to the skin.

The depiction of green colour under the skin strongly suggests solanine build-up in potatoes, although each process can occur without the other. A bitter taste in a potato is another, potentially more reliable indicator of toxicity. Because of the bitter taste and appearance of such potatoes, solanine poisoning is rare outside conditions of food shortage. The symptoms are mainly vomiting and diarrhea, and the condition may be misdiagnosed as gastroenteritis. Most potato poisoning victims recover fully, although fatalities are known, especially when victims are undernourished or do not receive suitable treatment Fatalities are also known from solanine poisoning from other plants in the nightshade family, such as the berries of *Solanum dulcamara* (woody nightshade). The United States National Institutes of Health's information on solanine advised against eating of potatoes that are green below the skin.

Deep frying potatoes at 170°C is known to effectively lower glycoalkaloid levels (because they move into the frying fat), whereas microwaving is only somewhat effective, freeze drying or dehydration has little effect, and boiling has no effect.

Solanine in tomatoes:

The California Poison Control System, have claimed that tomatoes and tomato leaves contain solanine. However, Dr. Mendel Friedman of the federal Department of Agriculture contradicts this claim when stating that tomatine, a relatively benign alkaloid, is the tomato alkaloid while solanine is found in potatoes. Food science writer Harold McGee has found scant evidence for tomato toxicity in the medical and veterinary literature.

Other uses of solanine:

Solanine has fungicidal and pesticidal properties, and so solanine hydrochloride (a salt of solanine) has been used as a commercial pesticide, but never on a large scale. Solanine has sedative and anticonvulsant properties, and has been used as a treatment for asthma, as well as for cough and cold. However, its effectiveness for either use is questionable.

Chapter 6

HYPERYTENSIVE EFFECT OF ORAL ADMINISTRATION OF AQUEOUS EXTRACT OF *SOLANUM TORVUM* FRUITS IN L-NAME TREATED RATS

EFFECT OF TREATMENT ON THE ORGAN WEIGHTS

Four weeks treatment with l-NAME (40 mg/kg/day) caused cardiac hypertrophy and increase in the cardiac (328.9±7.0 vs 290.8±6.8 in control) and left ventricular (245.2±5.0 vs 213.0±5.4in control) mass index. This effect was increased by the co- administration of AEST or l-arginine while captopril prevented its effects. AEST administrated alone did not affect this parameter. None of the treatment significantly affects the liver weight while the concomitant administration of l-NAME and l-arginine significantly increased the relative kidney weight (8.2±0.8) compared tocontrol (6.4 ±0.1).

EFFECT OF TREATMENT ON THE VASCULAR REACTIVITY TO NORADRENALINAND CARBACHOL

Rat aortic rings were pretreated and then contracted with 1M of noradrenalin (NA) concentration-dependently and then relaxed by carbachol (10−9 to 10−5 M) with an *E* max of 81.1±4.0% on organs from control animals. Chronic treatment with l-NAME significantly ($p < 0.001$) reduced this *E* max up to 50.3±4.3%. The addition of either l-arginine or AEST to l-

NAME treatment did not reduce the hyposensitivity induced by l-NAME while captopril rather aggravated it by reducing the carbachol *E*max up to 35.1±8.8%. In animals that were only treated with AEST, the effect of carbachol was significantly reduced as compared to control (*E*max 66.9±2.8%) but was still significantly high compared to l-NAME treated group.

NA produced a concentration-dependent contraction of aortic ring with no considerable difference between control and groups treated with l-NAME and l-NAME+ l-arginine. Aortic rings from rats treated with l-NAME+ AEST showed a remarkable reduce sensitivity to NA as compared to both control ($p < 0.05$) and l-NAME treated group ($p < 0.001$) while aortic rings from rats treated only with Vascular reactivity to carbachol (A) or noradrenalin (B) of aortas fromrats submitted chronically for 4weeks to different treatments. Results in panel B are expressed as a percentage of maximal contraction induced by KCl (60 mM). N=6; ap< 0.05, cp < 0.001 was significantly different compared to control; #p< 0.05, $$p$ < 0.01 which was significantly different compared to l-NAME. AEST showed a very high reactivity, increasing the *E*max of NA to 127.0±7.2 compared to 59.7±4.3 in control and 71.6±5.3 in l- NAME group.

EFFECT OF TREATMENT ON RENAL FUNCTION

Before treatment, urinary volume and kalliuresis were not significantly different between groups while natriuresis was significantly different between control, AEST and l-NAME+ captopril group. For this reason, we then do two different statistical analyses: intergroup comparing each group to control and intragroup comparing values of the same group at different time of treatment to the value obtained before treatment. Only animals from group treated with l-NAME+ AEST or l- NAME+ captopril showed significant increase in urinary volume both compared to control and their respective initial value. l-NAME given alone reduced the natriuresis by 32.5% (from 2.5±0.3 to1.7±0.2 mEq/kg/24 h) but the difference was not significant. Compared to the value before treatment, only AEST + l-NAME treatment significantly affected the kaliuresis. AEST given alone, acutely and chronically reduced the sodium excretion and the Na+/K+ ratio while l-NAME+ captopril increased it.

EFFECTS OF *SOLANUM TORVUM* ON NORADRENALIN-PRECONTRACTED AORTA

In aortic rings with intact endothelium, precontracted with NA (10−6 M), addition of AEST in the incubation medium induced a concentration-dependent additive contraction. At the extract concentration of 1000_g/ml, the contraction increased by about 30%. In endothelium denuded rings, extract concentrations ranging from 10 to 300_g/ml induced a slight contraction while the concentration of 1000_g/ml instead induced relaxation.

EFFECTS OF *SOLANUM TORVUM* ON THE BASAL TONE OF AORTA RINGS

To further investigate the contractile effect of AEST, its effects were tested on the basal tone of aortic rings. AEST induced a concentration dependent contraction of both intact and endothelium denuded aortic rings. At the concentration of 1000_g/ml, the effect of AEST was significantly high on intact aorta. Effects of the aqueous extract of *solanum torvum* fruits on the untreated rat aorta precontracted with noradrenalin. Contractions are expressed as percentage of the initial contraction induced by noradrenalin (10−6 M). N=6, c$p < 0.001$ compared to the extract effect on aorta with endothelium tic ring than on endothelium denuded organ and even than that of NA (10−6 M). Prazosin (10−6 M) and verapamil (10−5 M) significantly ($p < 0.001$) reduced the effects of AEST at the concentrations of 100 and 300_g/ml but failed to significantly affect the Emax of the extract obtained at the concentration of 1000_g/ml concentration of carbachol (10−5 M) that reduced the contraction induced by NA (10−6 M) by 78.9% but could not relax the contraction induced by AEST while the effect of sodium nitroprusside (10−4 M) was similar in both contractions.

Chapter 7

Cloning and Characterization of a PGIP- Encoding Gene from *Solanum torvum*

Polygalacturonase-inhibiting proteins (PGIPs) have been shown to involved in preventing the invasion of fungus such as *V. dahliae*. Cloning genes encoding PGIPs is quite valuable for plant resistance breeding to Verticillium wilt. For this purpose, *S. torvum* was grown in the greenhouse for 16 hours under light and 8 hours under darkness (25°C). Roots were collected and frozen in liquid nitrogen immediately and stored at -70°C for cloning of StPGIP. Leaves, stems and roots were collected and frozen for analysis of StPGIP expression in different tissues.A cDNA encoding the polygala-acturonase-inhibiting protein was isolated from *Solanum torvum* by RT-PCR and RACE, designated StPGIP (accession no. FJ943498). Analysis of StPGIP expression at the mRNA level by RT-PCR showed that the gene was expressed in all organs and could be induced to increase expression by *V. erticillium dahlia* infection.

Isolation of Full-length cDNA of StPGIP

By using RT-PCR and RACE, a full-length cDNA of *PGIP* was isolated from *S. torvum*. The cloning stated by amplifying about 750 base pair fragment of the conserved region, followed by generation of 3′- and 5′-fragmentsin a range of 750 and 350 base pairs, respectively. Based on the assembled sequence, the full-length cDNA of1097 base pairs was obtained which contained an open reading frame of 990 base pairs encoding a putative protein of 329 amino acids with a molecular weight of 36.9 kilo-Daltons and atheoretical *p*I of 8.68. Alignment of sequence indicated that the cDNA clone and its deductive encoding protein had high identity with *S. lycopersicum PGIP* and its protein (87 and 86% respectively) and *Daucus carota PGIP* and its protein (65 and 79% respectively). The cDNA sequence was designated StPGIP and has been deposited in GenBank database with accession no.FJ943498. Analysis of the deduced amino acid sequence of StPGIP. Analysis on protein structure was performed by SignalP, TargetP, SOSUI tool, and motif_scan tool, respectively. StPGIP protein contained a signal peptide domain, a secretory site, 9 N-glycosylation sites, 9 phosphorylation sites, and 10 leucine-rich repeats (LRR), which includes 3 plant-specific LRR profiles, 1 typical LRR profile, 1 LRR N-terminal domain, and 5 LRR. The structure is in consistence with PGIPs isolated from other plants (Yao *et al.*, 1999; Dou *et al.*, 2003; Ahsan *et al.*, 2005).

EXPRESSION OF STPGIP

The expression of StPGIP was analyzed by semiquantitative RT-PCR with cDNAs from the leaves (L), stems (S) and roots (R) of Solanum torvum. The result showed that the gene was expressed in all organs and higher in root. Further analysis on StPGIP expression in roots inoculated with V. dahlia showed that the expression of StPGIP increased with the prolongation of time after inoculation, reached the highest level at 12 hours and then decreased and kept at similar level compared with the control.

The ethanolic extract of Solanum torvum (100 and 300 mg/kg; p.o. for 6 weeks) effects the systolic blood pressure (SBP), vascular reactivity, serum glucose, triglycerides, cholesterol, insulin and uric acid in fructose-induced hypertension. The effect of ethanolic extract of Solanum torvum (100 and 300 mg/kg; p.o. for 6 weeks) on fructose (10%) induced rise in blood pressure was tested by invasive and non-invasive and the biochemical parameters were studied using standard kits. Ethanolic extract of Solanum torvum reduced systolic blood pressure, vascular reactivity changes to catecholamines and reversed the metabolic alterations induced by fructose. In conclusion, ethanolic extract of Solanum torvum could prevent the development of high blood pressure induced by a diet rich in fructose probably by reversing the metabolic alterations induced by fructose.

A b-glucosidase (torvosidase) was homogenously purified from the young leaves of Solanum torvum. The enzyme was highly specific for cleavage of the glucose unit attached to the C-26 hydroxyl of furostanol glycosides from the same plant, namely torvosides A and H. Purified torvosidase is a with a native molecular weight of 87 kDa was used as a monomeric glycoprotein, by gel filtration and a pI of 8.8 by native agarose IEF. Optimum pH of the enzyme for p-nitrophenyl-b-glucoside and torvoside H was 5.0. According to Kinetic studies Km values for torvoside A (0.063 mM) and torvoside H (0.068 mM) were much lower than those for synthetic substrates, pNP-b-glucoside (1.03 mM) and 4 methylumbelliferyl-b-glucoside (0.78 mM). The enzyme showed strict specificity for the b-D-glucosyl bond when tested for glycone specificity. Torvosidase hydrolyses only torvosides and dalcochinin-80-b-glucoside, which is the natural substrate of Thai rosewood b-glucosidase, but does not hydrolyse other natural substrates of the GH1 b-glucosidases or of the GH3 b-glucosidase families. Torvosidase also hydrolyses C5–C10 alkyl-b-glucosides, with a rate of hydrolysis increasing with longer alkyl chain length. The internal peptide sequence of Solanum b-glucosidase shows high similarity to the sequences of family GH3 glycosyl hydrolases.

Phylogenetic Analysis

A phylogenetic tree was constructed to determine the phylogenetic relationship of StPGIP with PGIPs after aligned with other PGIPs from 17 plant species downloaded from GenBank. This dendrogram revealed that the StPGIP protein was closer to PGIPs of S. lycopersicum and Dacuscarota. But they did not belong to the same cluster.

Chapter 8

PHARMACOLOGICAL PROPERTIES

Solanum torvumhas both a sedative and a diuretic therapeutic effect. The leaves are used as a haemostatic. Phytochemical studies indicated that fruits of this species have a good concentrations of various alkaloids, flavonoids, saponins, tannins, and glycosides sufficient to have pharmacological effects. Therefore, fruits are not only used for nutritive purposes but also fruit decoctions are effective for cough ailments and are considered effective medicine for cases of liver and spleen enlargement (Kala, 2005). The ripened fruits are used in the preparation of tonic and haemopoietic agents and also for the treatment of pains (Kala, 2005). The methanolic extracts of fruits and leaves were reported to have antimicrobial activities against human and animal clinical isolates (Chah et al., 2000). An antiviral isoflavonoid sulfate and various steroidal glycosides were also isolated from fruits (Arthan et al., 2002). However, S. torvum exhibited some antioxidant activity and DNA-repair capability in oxidative DNA damage caused by free radicals (Abas et al., 2006). Recently, a novel protein was isolated from water extracts of seed that has been proved to be an effective antioxidant. Sivapriya and Srinivas (Sivapriya et al., 2007) concluded that these proteins are effective even at low doses when compared to well-known standard synthetic antioxidants. Aqueous extracts from the various parts of S. torvum exhibit potential anti-inflammatory and analgesic properties (Ndebia et al., 2007). This herb is commonly used in China and considered effective in the treatment of blood stasis, menstruation, edema pain and coughs.

ANTIMICROBIAL ACTIVITY

Methanolic extracts from sun-dried fruit of S. torvum were found to have effective antimicrobial activity against human and animal clinical isolates. Biochemical analyses of methanolic extracts indicate the presence of alkaloids, flavonoids, saponins, tannins, and glycosides. Chahet al., in 2000 used methanolic extracts against bacteria (Actinomycespyogenes, Bacillus subtilis, Escherichia coli, Klebsiellapneumoniaed, Pseudomonas aeruginosa,, Salmonella typhimuriumd, Staphylococcus aureus, Streptococcus pyogenes) fungi(Aspergillusniger, Candida albicans)isolated from the clinical samples of humans and animals to evaluate antimicrobial activity by the disc diffusion method. The study concluded that methanolic extracts have significant growth-inhibiting activity against bacteria commonly associated with pyogenic infections (Fig 14). The minimum concentration that inhibits bacterial and fungal activity were 0.3125mg/ml and 1.25mg/ml respectively. However, growth of Bacillus cereus, Bacillus subtilis, Candida albicans, Escherichia coli, Pseudomonas aeruginosa and Staphylococcus aureuswas monitored in the presence of methanolic leaf extracts of S. torvum by Wiart et al., in 2004 ; S. torvum leaf extracts showed the highest activity against B. cereus. Methanolic extracts from fruit and leaves have been described as potentially good sources of antimicrobial agents (David *et al.,* 1998). These in-vitro studies collectively support the ethnobotanical use of *S. torvum*. Nevertheless, methanolic extracts from various parts (leaves root and stem) were investigated in the Biomphalariaglabrataassay (Tania et al., 2007). It was discovered that S.torvum extracts are active in brine shrimp bioassay but were inactive against the mollusk Biomphalariaglabrata leading to the conclusion that those chemical compounds responsible for the inhibition of microbial activities are present in methanolic extracts of leaves and fruit rather than in the whole-plant extracts. Therefore, leaf and fruit methanolic extraction can be further utilized for the identification of the active compounds.

ANTI-ULCEROGENIC ACTIVITY

Ulcers are basically discontinuities of the skin or break in the skin, or can be on inner layers of the skin, which stops it from continuing its normal protective functions. This breakdown of the skin can occur at any place in the body leading to various types of ulcer such as peptic, corneal, venous and genital ulcers. *S. torvum* is reported to have anti -ulcerogenic effect. Nguelefacka *et al.,* (2008) investigated anti-ulcerogenic properties of aqueous and methanolic extracts from leaves of *S. torvum* in rats. They induced gastric ulceration by HCl/ethanol, indomethacin, pylorus ligation, and stress. Various doses of aqueous and methanolic extracts at the rate of 250, 500 and 750 mg/kg were tested. Methanolic extracts were fractionated into seven different fractions (A–G) through silica gel column chromatography. These fractions were tested orally at the dose of 100 mg/kg against HCl/ethanol-induced ulceration. Methanolic extracts at the dose of 750 mg/kg produced 98.12, 99.16, 98.70 and 96.03% inhibition when gastric ulcerations were induced by HCl/ethanol, indomethacin, pylorus ligation and stress, respectively (Table 2). Aqueous extracts at the same dose produced 96.55, 96.86, 98.63 and 98.63% inhibition on ulcerations induced respectively by HCl/ethanol, indomethacin, pylorus ligation and stress. All the fractions of the methanolic extracts significantly inhibited ulcer formation. Fractions that contained flavonoids and triterpenes were the most active and exhibited an inhibitory percentage of 84.74 (Table 2). From these results it was obvious that aqueous and methanolic extractions promote production of mucus and reduced gastric-acid secretion. Therefore, although it has effective control of gastric ulceration, *S. torvumextracts* need to be explored for their effectiveness against peptic, corneal, venous and genital ulcers.

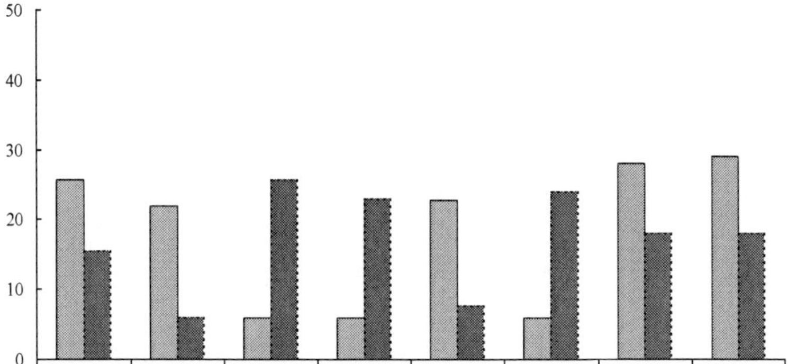

	Actinomyces pyogenesd	Bacillus subtilis	Escherichia coli	Klebsiella pneumoniaed	Pseudomonas aeruginosa	Salmonella typhimuriumd	Staphylococcu s aureus	Streptococcus pyogenesd
MeOH extract 80mg/ml	25.7	22	6	6	22.8	6	28	29
Standard2mg/ml	15.5	6	25.6	23	7.7	24	18	18

Figure 14. Antibacterial activity of the methanolic extract of *Solanum torvum* fruit.

Minimum inhibitor concentration of MeOH extract is 0.3125 mg/ml.

Table 2. Effect of *Solanum torvum* extracts on gastric lesions induced by four methods

S. No	Mode for induction of ulcer	Methanolic Extracts		Aqueous Extract	
		Conc. mg/kg	Inhibition percentage	Conc. mg/kg	Inhibition percentage
1	Pylorus ligation	250	47.62	250	50.56
2		500	78.75	500	78.62
3		750	98.70	750	98.63
	indomethacin	250	44.60	250	41.83
		500	80.33	500	77.82
		750	99.16	750	96.86
	HCl/ethanol	250	41.22	250	36.87
		500	79.68	500	78.39
		750	98.12	750	96.55
	Cold-restraint stress	250	33.51	250	38.97
		500	73.20	500	73.61
		750	96.03	750	95.81

ANTIVIRAL ACTIVITY

Herpes simplex virus (HSV) is responsible for human infections in the orofacial region (HSV-1) and in the genital region (HSV-2) (Travis, 2002). Type 1 (HSV-1) of Herpes simplex virus was tested against methanolic fruit extracts of *S. torvum*. HSV-1 was maintained in the Vero cell line (kidney fibroblast of an African green monkey), which was cultured in Eagle's minimum essential medium (MEM) with addition of heat-inactivated fetal bovine serum (FBS) (10%) and antibiotics. The test samples were put into wells of a microtiter plate at final concentrations ranging from 20 to 50 mg/ml. The viral HSV-1 (30 PFU) was added into the 96-well plate, followed by plating of Vero cells (1105 cells/ml); the final volume was 200 ml. After incubation at 37 ºC for 72 hours, under 5% CO_2 atmosphere, cells were fixed and stained, and optical density measured at 510 nm. Under these screening conditions, the reference compound, Acyclovir, typically exhibited an IC50 of 2-5 mg/ml for HSV-1. Torvanol A, torvoside H, and solasoninealso inhibited the expression of herpes simplex virus-1 (HSV-1), and the activity may relate to the entering of glycosides into the viral capsules (Arthan *et al.,* 2002).

EFFECT ON SYSTOLIC BLOOD PRESSURE

High blood pressure is generally considered to be induced by a diet rich in fructose. Ethanolic extracts of *S. torvum* acts by preventing and reversing the development of hyperinsulinemia to control the rise in systolic blood pressure (Mohan*et al.,* 2009). The Mohan group investigated the effect of *S. torvum* on blood pressure and metabolic alterations in the fructose hypertension condition generated in rats. They found that the hypertensive conditions were induced in male Wistar rats (150–200 g) by a high-fructose diet (fructose 10%, w/v) *ad libitum*for six weeks. For measurements of systolic blood pressure, non-invasive (indirect) and invasive (direct) methods were used. Ethanolic extractions of *S. torvum* were demonstrated to be effective in preventing high systolic blood pressure (Fig 15). Nguelefack*et al.,* (2008, 2009) also supported the findings of Mohan *et al.,* (2009). They orally administered aqueous extracts of *S. torvum* fruit (AEST) to investigate effectively of AEST in hypertensive conditions. Rats were given AEST (200 mg/kg/day, *p.o.*) solely or concomitantly with l-NAME (40 mg/kg/day, *p.o.*) for 30 consecutive days. It was found that AEST induced neither mortality nor visible signs of toxicity. When given solely or in co-administration with.

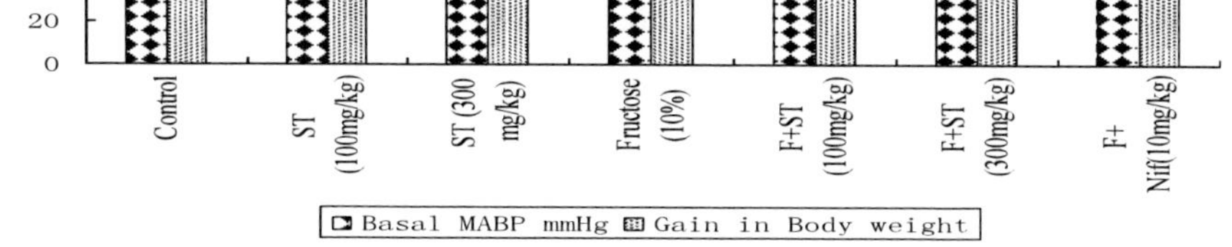

Figure 15. Effect of *Solanum torvum*(100 mg/kg and 300 mg/kg, p.o., for 6 weeks) on mean arterial blood pressure and change in body weight in fructose (10%) induced hypertensive rats.

However l-NAME put the rats in hypertensive conditions, whereas AEST increased the sensitivity to noradrenalin. In normotensive and significantly reduced it in hypertensive animals. AEST significantly increased urinary volume and sodium excretion in l-NAME treated animals while reducing the sodium excretion in normotensive to reduce hypertensive conditions.

Anti-platelet Aggregation Activity

Haemostatic properties of *S. torvum* impart to it an anti-platelet aggregation effect (Henty, 1973). The anti-platelet aggregation activity of aqueous extracts of *S. torvum* (AES) was evaluated in vitro on platelet aggregation initiated by thrombin and ADP (Nguelefacka *et al.,* 2008). Results showed that anti-platelet aggregation activity is concentration dependent. At 2 mg/ml, AES reduced the amplitude of the aggregation signal induced by thrombin from 9.27 cm to 4.03 cm, representing an inhibition of 55.27%. This effect was significantly higher than that of the lower concentrations 0.5 and 1 mg/ml. Similarly, AES also exerted significant concentration-dependent inhibitory effects on aggregation triggered by ADP. The effect in terms of inhibitory percentage was 31.63%, 47.07%and 56.40% at concentrations of 0.5, 1 and 2 mg/ml respectively.

Anti-oxidant Activity

Phenolic compounds extracted from different parts of *S. torvum* exhibited anti-oxidant activity (Loganayaki *et al.,* 2010). Chloroform, acetone, and methanol extracts of leaves and fruit were explored for their in-vitro antioxidant activity using ferric reducing antioxidant power, 2,2- diphenyl-1-picryl-hydrazyl (DPPH), ABTS•+, iron chelation, and anti-hemolytic activity. Significantly higher concentrations of phenol were recorded for chloroform extracts (Table 3). Of note was the fact that the in-vitro antioxidant activity was shown to be highly dependent on total phenolic content ($p<0.01$). The DPPH and 2, 2' azinobis (3- ethyl benzothiozoline -6- sulfonic acid) diammonium salt (ABTS) cation radical scavenging activities were well proved with the ferric reducing antioxidant capacity of the extracts. However, peroxide clearing studies indicated that the aqueous extracts of *S. torvum* fruit had anti-oxidant activities (Re *et al.,* 1999). This species also exhibited some percentage of antioxidant activity and DNA-repair capability on oxidative DNA damage caused by free radicals (Abas *et al.,* 2006).

Table 3. Extract yield, total phenolic content in various solvent extracts from *S. torvum*and DPPH, antihemolytic and FRAP activities

Samples	Extract Yield (%)	Total Phenolics (gTAF/100g extract)	DPPH radical scavenging activity 1C$_{50}$ DPPH (μg)	Antihemolytic activity (%)	Ferric-reducing/antioxidant power assay (FRAP)
S. torvum leaf Chloroform extract	34.4	2.90.1	30	71.2	537±40.8
S. torvum leaf Methanol extract	7.8	2.10.1	55	54.9	46.5±02. 9
S. torvum leaf Acetone extract	4.1	1.30.1	40	56.3	143.9±13. 2
S. torvum fruit Chloroform extract	68.2	8.50.4	28	92.6	552.8±28. 1
S. torvum fruit Methanol extract	3.8	0.80.0	50	58.4	427.3±69. 3
S. torvum fruit Acetone extract	2	0.40.0	58	58.9	206.1±16. 8

CYTOTOXIC EFFECT

Among the chemical constituents of *S. torvum,* steroidal lactone saponins and spirostanol glycosides are important for cytotoxicity. Lu *et al.,*(2009) isolated new steroidal lactone saponins and spirostanol glycosides from ethanolic extracts of aerial parts. All the four compounds (Solanolactoside A, Solanolactoside B, Trovoside M and Trovoside N) were evaluated for cytotoxic effects *in vitro* against a panel of human cancer cell lines: MGC-803 (Human gastric cancer cell line), HepG2 (Human hepatocellular liver carcinoma cell line), A549 (human lung adenocarcinoma cell line) and MCF-7 (Human breast adenocarcinoma cell line). Cis-diamminedichloroplatinum (CDDP, Sigma) was used as a positive control. The dose concentration was maintained at 5mg/ml. The effects of two compounds (Solanolactosdie A, Solanolactoside B) were not significant. However, the other two compounds Torvoside M and Torvoside N had significant cytotoxicity.

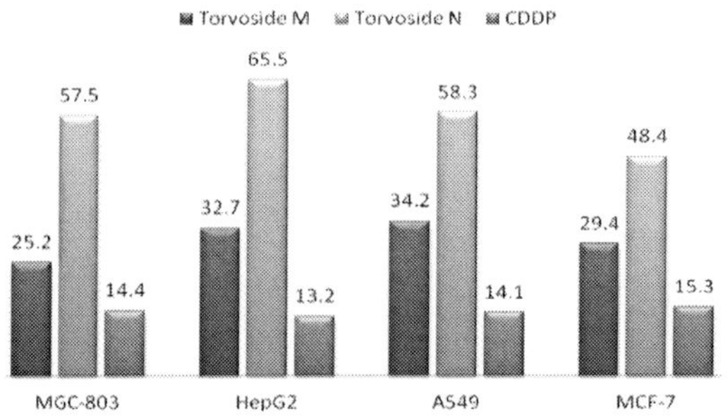

Figure 16. Cytotoxicity of saponins extracted from ethanol extract of aerial parts of *Solanum torvum.*

ANALGESIC AND ANTI-INFLAMMATORY EFFECTS

S. torvum is amongst the important medicinal species used as analgesic and anti-inflammatory agents in different traditional medicinal systems (Ndebia *et al.,* 2007). The analgesic and anti-inflammatory activity of *S. torvum* was evaluated for chemical and mechanical stimuli. Abdominal writhing and paw edema were induced in rats by using 1% acetic acid (1 ml/100 g body weight) and 0.05 ml of a solution of 1% sterile carrageenan in saline, respectively. For the treatment of 1% acetic acid (1 ml/100 g body weight) induced abdominal writhing, aqueous extracts of *S. torvum* were utilized along with three other painkillers. The aqueous leaf extracts of *S. torvum* significantly inhibited the pain (Figure 17)

Table 4. Anti-inflammatory activities of aqueous extracts of *S. torvum* in carrageenan-induced hind paw edema (in terms of % inhibition)

Treatments	Dose (mg/kg)	% inhibition			
		1H	2H	3H	4H
Indomethacin	10	43	62**	82**	85**
S. torvum	300	34	31	39**	47**
S. torvum	600	11	4	41	49**

Paw edema induced by the 0.05 ml of a solution of 1% sterile carrageenan was treated with indomethacin (10mg/kg), *S. torvum* (300mg/kg) and *S. torvum* (600mg/kg). Extracts of *S. torvum* (300mg/kg) and *S. torvum* (600mg/kg) significantly inhibited the paw edema although the lower dose

300mg/kg worked more effectively in a shorter time period compared with the high dose of 600mg/kg.

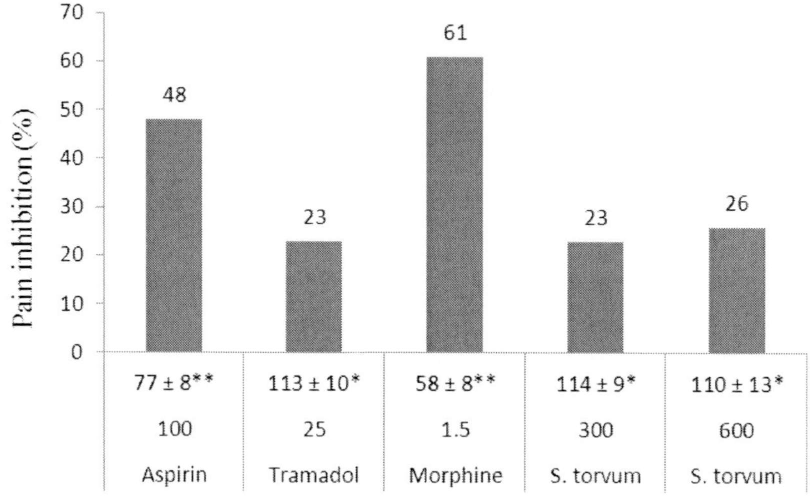

Figure 17. Effect of the aqueous leave extracts of *S. torvum* on acetic acid-induced pain.

EFFECTS ON SMOOTH MUSCLES

Solasonine inhibited cat isolated heart contraction and contraction of the isolated guinea pig ileum and cat trachea induced by acetylcholine. It also had a contracting effect on isolated rabbit aural blood vessels, and could induce contraction and spontaneous activities of isolated rat uterus (Basu and Lahiri, 1997).

Chapter 9

Plant Specimen, Extraction and Isolation

Crushed fruits of *S.torvum* were macerated in MeOH for 2 days at room temperature. The extract was evaporated and dissolved in 70% aqueous MeOH, which was sequentially extracted with hexane and EtOAc. Twenty fractions were collected and evaporated to dryness in vacuo. Fractions 15–20 containing torvanol A which was subsequently purified by preparative HPLC using H2O: MeOH (90:10, v/v) as eluent, yielding torvanol A (25 mg). Fractions 3–6 obtaining from Sephadex LH-20 column were further purified by MPLC using H_2O: MeOH (60:40, v/v) as eluent, to yield torvosideA (2.1 g) and torvosideH (0.2 g).

Bioassay

The cytotoxic assay involved the use of colorimetric method (Skehan *et al.,* 1990). Antiviral activity was also assessed by employing the colorimetric method (Skehan *et al.,* 1990). Herpes simplex virus type 1 (HSV-1) was maintained in the Vero cell line which was cultured in Eagle's minimum essential medium (MEM) with addition of heat-inactivated fetal bovine serum (FBS) (10%) and antibiotics. After incubation at 37 C for 72 hours, the cells were fixed and stained fewer than 5% of CO2 atmosphere and optical density was measured at 510 nm. The reference compound, Acyclovir, typically exhibited the antiviral HSV-1 with the IC50 of 2-5 mg/ml under our screening conditions.

Extraction and Isolation

Dried aerial parts of *S. torvum* between 20 kg were extracted three times with 60% EtOH (vol. 70 L), and the extraction was carried out for 3hours each time. After concentrated under residue pressure, the water solution was subjected to a macroporous resin D101 column eluted with H2O/EtOH to give 5 fractions (1–5), respectively. On silica gel fraction 4 (251 g) was separated by column chromatograph to further give 9 fractions (4.1–4.9). Fraction 4.3 about 20 g was chromatographed again on silica gel and ODS column chromatography eluted with H2O/MeOH to give **1** (15 mg), **3**(98 mg) and **4** (40 mg), respectively. Fraction 4.4 about 29 g was submitted to MCI gel CC (H2O/MeOH, 40:60, v/v) and further to silica gel CC (CHCl3/MeOH/H2O, 9:1:0.2, v/v/v) to afford **2** (8 mg).

Acid hydrolysis of 1–4 and Determination of Absolute Configuration of Monosaccharides

A solution of **1–4** each 4 mg in MeOH was treated separately with 5% H_2SO_4 at 90◦C for 2hours. After added H_2O about 5 ml, the reaction mixture was evaporated to 10 ml under vacuum for removing of MeOH. The hydrolyzed mixture was neutralized with ion exchange resin and condensed to dryness under reduced pressure. The presence of the monosaccharides in the hydrolysates of each compound was confirmed by TLC comparison with authentic sugars, d-quinovose was detected from **1**, **2**; l-rhamnose was detected from **1**; d-xylose was detected from **2**; d-glucose was detected from **3**, **4**. Each remaining aqueous layer was concentrated to dryness to give a residue and dissolved in pyridine (2 ml), and then l-cysteine methyl ester hydrochloride (2 mg) was added to the solution. The mixture was heated at 60 ◦C for an hour, then added trimethylchlorosilane (0.5 ml) and the solution was concentrated to dryness and taken up in water (1 ml), followed by extraction with n-hexane (1 ml ×3). The supernatant was subjected to GC/MS analysis under the following conditions: Varian CP- 3800 Gas Chromatograph equipped with a Saturn 2200 Mass detector (detection temperature 220 ◦C). Column: CP-sil 5 CB capillary column (30 m, 0.25mm i.d., 0.25_m). Column temperature: 150–260 ◦C with the rate of 8 ◦C/min, and the carrier gas was He (0.8 ml/min), split ratio 1/10, injection temperature:250 ◦C; injection volume: 0.5_l. The absolute configurations of the monosaccharides were confirmed to be d-quinovose, l-rhamnose, d-xylose and d-glucose by comparison of the retention times of monosaccharide derivatives with those of standard samples: d-quinovose (12.51 min), l-rhamnose (12.67 min), d-xylose (11.89 min) and d-glucose (14.01 min), respectively.

CYTOTOXIC ASSAY

Cytotoxicities of the saponins were tested against a panel of mammalian cells that included four human cancer cell lines (MGC-803 (Human gastric cancer cell line), HepG2 (Human hepatocellular liver carcinoma cell line), A549 (human lung adenocarcinoma cell line) and MCF-7 (Human breast adenocarcinoma cell line)). The cells were obtained from the Cell Bank of the Shanghai Institute of Cell Biology. Compounds **1–4** were prepared as 5mg/ml top stocks, dissolved in DMSO, and stored at 4 ◦C, protected from light. The tumor cells were routinely cultivated at 37 ◦C in an atmosphere of 5% CO2 inRPMI-1640 medium (GibcoTM, Invitrogen Corporation) supplemented with 10% fetal calf serum (Sijiqing Biomaterial Co.,Hangzhou, China) and sub-cultured twice weekly to maintain continuous logarithmic growth. The cells (5×103 cells/well) were cultured into 96-well microtiter plates and allowed to adhere for 24hours before the drugs were introduced. Serial drug dilutions were added to each culture. At the time of drug addiction (parallel triplicate wells were set) following 48hours of culture, MTT solution (10_l; 5mg/ml) was added to each well. After a further 4hours of incubation, DMSO (150_l) was added to each well, and the form azan crystals in each well were dissolved by stirring with a pipette. Absorbance was read on a systems plate reader (Bio-Rad 680) at 570nm as a measure of cell viability.IC50 (the concentration of the test compound that caused a growth inhibition of 50% after 48hours exposure of the cells)was calculated from the dose curves generated by plotting percent viable cells versus the test concentration on a logarithmic scale using Microsoft Excel. Cis-diamminedichloroplatinum (CDDP, Sigma) was used as a positive control.

TESTED MATERIAL

MeOH extract of sun-dried fruits _yield: 8.2%. Phytochemical screening w5x gave positive tests for alkaloids, flavonoids, saponins, tannins and glycosides.

STUDIED ACTIVITY

Antimicrobial activity by the disc diffusion method and determination of the minimum inhibitory concentration MIC by broth dilution method w6x.

USED MICROORGANISMS

They microorganism were isolated from clinical specimens of human and animal submitted to the diagnostic units of the Department of Microbiology and Department of Veterinary Pathology and Microbiology, University of Nigeria, Nsukka.

APPARATUS AND REAGENT

NMR spectra were recorded on Bruker DRX-500 spectrometers (1H NMR at 500 MHz; 13C NMR at 125 MHz) using pyridine-d5 or CDCl3 as solvent with tetramethylsilane (TMS) as internal reference. ESI-MS was performed on an Agilent 1100 Series LC/MSD Trap mass spectrometer. Silica gel (45-75 μm, Qingdao Marine Chemical Co., Ltd., China), MCI gel CHP-20P (Mitsubishi Kasei Co.) and Rp-C18 (40-63 μm, Fuji) were used for column chromatography.

Chapter 10

SOLANACEOUS STEROIDAL GLYCOALKALOIDS AND POISONING

Ingestion of immature, environmentally stressed, or cultivar-specific *Solanum* species (particularly the potato) has been previously associated with gastrointestinal and neurological symptoms caused by solanaceous steroidal glycoalkaloids (SGAs). Two geographically, temporally disparate outbreaks of poisoning by (*Solanum torvum*– Solanaceae) and on detection of alkaloids absent in non-toxic berries. Five family members in New York City participated in a traditional evening meal containing Jamaican susumber berries.

All those consuming berries were symptomatized the following morning with varying degrees of gastrointestinal distress, dizziness, slurred speech, cranial nerve deficits, and ataxia. The most seriously afflicted patient developed hypertension, confusion, proximal upper extremity weakness, and hypercapnic respiratory failure requiring prolonged mechanical ventilation. A separate cohort of six patients in Toronto ate unripe Jamaican susumber berries. They presented 14 hours post-ingestion with varying degrees of diarrhea, weakness, facial paralysis, slurred speech, ataxia, early hypertension, and proximal weakness. Two patients had ventilatory decompensation; one required intubation. Poisonous berries appeared indistinguishable from non-toxic varieties. We isolated solasonine, larger amounts of solamargine, and other steroidal glycoalkaloids in the toxic berry strains. *S. torvum* poisoning can produce significant neurological and gastrointestinal effects which appear to be mediated by SGAs present in the berries.

SGAs are composed of a carbohydrate moiety (typically chacotriose or solatriose) linked to a steroidal glycone. Hydrolysis may occur in the human stomach via gut bacteria. Synthesized from cholesterol, SGAs confer insect, fungi, and nematode resistance (Morris and Lee, 1984). SGAs appear to act synergistically but several different types may often be found together in the same plant. Because of the difficulties in differentiating SGAs, early poisonings by Solanum spp. were attributed to ''solanine.'' A large outbreak afflicted 78 schoolchildren in England who consumed left-over potatoes intended for disposal which contained 25–30 mg solanine/100 g peeled potato (McMillan and Thompson, 1979). After 8–10 hours delay, gastrointestinal signs (vomiting, diarrhea, or general abdominal pain) predominated; the more seriously afflicted had associated CNS symptoms (confusion, delirium, hallucinations, restlessness, or coma).

In Canada 61/109 students were afflicted with rapid onset vomiting, abdominal pain, and diarrhea following ingestion of green potatoes containing 49.4 mg solanine/100 g potato (Anonymous, 1984). Analysis of *S. dulcamara* berries consumed by a 4-year-old girl who presented with tachycardia, mydriasis, rigidity, coma, and leucocytosis revealed solasodine and diogensin in unripe berries (Ceha *et al.,* 1997). Gastrointestinal toxicity predominates in SGA poisoning with a latency of 2–20 hours (Morris, 1859; Slanina, 1990). SGA concentrations are greater than 22 mg/100 g tuber produced mild to severe burning sensations in the mouth and throat (Jadhav *et al.,* 1981). Humans are presumably rarely affected due to the relatively poor gastrointestinal SGA bioavailability and consumption of domesticated *Solanum* species with lower SGA content.

However, SGAs are clearly absorbed: serum solanidine concentrations correlated with degree of potato consumption in 57 human volunteers (Harvey *et al.,* 1985). Peak concentrations of 7.7 ng/mL a-solanine and 14.4 ng /mL a-chaconine were detected in seven human volunteers consuming 0.4 and 0.6 mg/kg a-solanine and a-chaconine, respectively. Biological halflives were 10.7 hours for a-solanine and 19.1 hours for a-chaconine (Hellenas *et al.,* 1992). All but one volunteer complained of abnormal taste and light to severe nausea, one of the cases was in combination with diarrhea. In 14 adult volunteers given S.W. / Toxicon 52 (2008) 667–676 669 oral ascending doses of a-solanine and a-chaconine, 1.25 mg/kg total SGA produced peak concentrations at 6 hours of 8.0–11.8 ng/mL a-solanine and 6.3–9.7 ng/mL a-chaconine; no saturation of absorption was noted(Mensinga *et al.,* 2005). Emesis was noted at 4 h in one of two subjects at this dose. Humans intravenously administered H3-solanidine demonstrated erythrocyte uptake, and extremely low excretion rates (2% per day), indicating that LC/UV chromatograms at 205 nm of ''poisonous'' New

York berries (a), ''poisonous'' Toronto berries (b), and ''normal'' commercial berries (c). Locally obtained market berries were similar to (c).

There are numerous peaks present in the ''poisonous'' berries (a,b), particularly in the 3–6-min elution time range, which are absent from the ''normal'' berries (c). 670 S.W. Smith *et al.,*Toxicon 52 (2008) 667–676 significant SGA accumulations might occur (Claringbold *et al.,* 1980). Recovery of one of the aglycones (8) in one affected patient's urine clearly demonstrates that systemic. SGA absorption is possible in susumber berry ingestion.

Neurological and central nervous system effects of SGAs have been described in previous outbreaks. While investigating the extent of human plasma acetylcholinesterase inhibition by plant pesticide residues, it was unexpectedly discovered that plant SGAs themselves inhibited plasma cholinesterase (Orgell *et al.,* 1958). Rabbits given 20 or 30 mg/kg solanine IP demonstrated muscle weakness resembling partial paralysis, dilated pupils, respiratory depression and occasional convulsions (Patil *et al.,* 1972).

Both serum and erythrocyte anticholinesterases were inhibited by 68–100%. In vitro the SGAs a solanine and achaconine inhibit human recombinant acetylcholinesterase (AChE) by 50% at 14 and 17 mM (analogous to atracurium) and human plasma butyrylcholinesterase (BuChE) by 50% at concentrations of 0.17 and 0.066 mM (more potent than pancuronium) (McGehee *et al.,* 2000).

CONCLUSION

Solanum torvum is a pharmaceutically important member of the potato family. The species is included among the ingredients of various indigenous herbal medicines for treating a number of diseases. Moreover, anticancer phenolic compounds have also been isolated from fruit and leaves of this plant. The presence of various potentially important compounds justifies exploration of its medicinal qualities reinforced by its importance in indigenous herbal and conventional medicines. In addition, there is need to explore the local indigenous uses of this plant in different communities.

This review reveals that the phytochemistry of *S. torvum* will remain an important topic of pharmaceutical research. It has been extensively explored for compounds such as alkaloids, phenols, and steroidal saponins. Isoflavonoid sulfate and steroidal glycosides are important antiviral compounds and they have been isolated from the fruit. To date, steroidal glycosides (22, 3, 10, 41; 12), and long-chain hydrocarbons as well as steroids (Mohmood *et al.,* 1983) have been isolated and evaluated for their therapeutic effect.

Literature searches on *S. torvum* revealed that research on its biodiversity, conservation; genetic diversity, phylogeny, nucleotide sequence, chemotaxonomic variation, and allelopathy are yet to be conducted. These areas of research should be explored in the future.

RECOMMENDATIONS

SHORT-TERM GOALS

1. Genetic diversity within and among populations of *S. torvum* must be evaluated.
2. Chemotaxonomic studies must also be conducted to determine chemotypical variations in populations of *S. torvum*. Genus and family specific alkaloids, steroidal saponins, phenols etc. of *S. torvum* must be identified and used as a taxonomic tool in chemotaxonomy.
3. As *S. torvum* is always collected from the wild resources, there might be reduction in the population of this species. The only way to reduce the pressure on natural resources is to bring medicinally important plant species into cultivation. Therefore, it is important to explore the relationships between *S.torvum* with other species growing in its close vicinity. Allelopathic studies of *S. torvum* should also be conducted. On the basis of its richness in various chemical compounds, this species may also possess allelochemicals. Therefore, studies on the identification and isolation of dominant allelochemicals of *S. torvum* are required

Long-term Goals

The safety and efficacy of herbal drugs and their ingredients used for the treatment of various diseases must be thoroughly investigated and monitored independently.

REFERENCES

Abas, F., Kajis, N.H., Israf, D. A., Khozirah, S., &Umikalson, Y. 2006. Antioxidant and nitric oxide activities of selected Malay traditional vegetables. *Food Chem. 95:* 566–573.

Agra M.F, Bhattacharyya.1999Ethnomedicinal and phytochemical investigation on the *Solanum* species in the Northeast of Brazil. In Nee M, Symon DE, Lester RN, Jessop JP (eds.). *Solanaceae IV*. Kew: *Royal Botanic Gardens* p. 341-343.

Ahsan, N., Yoon, H. S.,& Jo, J. 2005. Molecular cloning of a bcpgip cdna from *Brassica campestris*and its expression to several stresses. *Plant Science, 169,* 1081-1089.

American board of veterinary toxicology.2001. An overview of cholecalciferol toxicosis. Http://Www.Abvt.Org/Tow.Htm. 6 P.And Collaborators). *Ann. Miss. Bot. Gard. 60(3),* 573-780.

Anonymous. 1984. Solanine food poisoning associated with a school lunch program alberta. *Can. Dis.* Wkly Rep*71,*10-18.

Anonymous. 2000. Census report of kurram agency. Population Census Organization, Printing Press Islamabad, Pakistan.Pp. 103.

Arthan, D., Svastia, J., Kittakoopb, P., Pittayakhachonwutb, D., Tanticharoenb, M.,& Thebtaranonth, Y. 2002. Antiviral isoflavonoid sulfate and steroidal glycosides from the fruits of *Solanum torvum. Phytochemistry.*459–463.

Badola, K.C., MohinderP., Bhanderi,H.C.S., & Pal, M.1993. Vegetative propagation of ranbaigan (*Solanum torvum*Sw.)By Rooting Branch Cuttings. *Indian Forester 119(12),* 1,027- 1,028.

Basu A& Lahiri SC. 1997. Some pharmacological actions of solasonine. *Ind J of ExpBio,* 15(4): 285-289.

Ceha, L.J., Presperin, C., Young, E., Allswede, M., &Erickson, T. 1997. Anticholinergic toxicity from nightshade berry poisoning responsive to Physostigmine.*J. Emerg. Med 15 (1)*, 65–69.

Chah,U. K.F., Mukob, K.N., & OboegbulemaS.I. 2000. Antimicrobial activity of methanolic extract of *Solanum torvum*fruit fitoterapia.187-189.

Claringbold, W.D., Few, J.D., Brace, C.J., &Renwick, J.H. 1980. The disposition and metabolism of [3h]-solasodine in man and syrian hamster. *J. Steroid Biochem 13 (8),* 889–895.

Chellaiah M., Muniappan A.,Nagappan R & Savarimuthu I.2006 Medicinal plants used by traditional healers in Kancheepuram District of Tamil Nadu, *Ind J of Ethnobiology and Ethnomedicine*, 2:43.

CPR Environmental Education Centre. 2001. Medicinal Plants. Http://Cpreec.Org/Edu/Medi-Pln.Htm. 4 P.

D'Arcy, W.G. 1973. Solanaceae.In flora of panama, part 9. (R.E. Woodson, R.W. Schery).

D'Arcy, W.G. 1974. *Solanum*and its close relatives in florida. *Ann. Miss. Bot. Gard.61(3),* 819-867.

David L.L., Alice M.C., Charles D.H., Barbara M., Claus M., Passreiter J.C., Omar I & Adewole L.O. 1998. Antimicrobial properties of Honduran medicinal plants.*Journal of Ethnopharmacology*, *63*(1):253-263.

Dou, D. L., Wang, B. S., Tang, Y. X., Wang, Z. X., &Sun, J. S. 2003. Cloning and sequence analysis of a gene encoding polygalacturonaseinhibiting protein from cotton. *Progress in Natural Science13,* 119-124.

Edmonds, J.M. 1977. Taxonomic studies on *Solanum*L. section *Solanum* (Maurella). *Bot. J. Linn. Sot.75,* 141-178.

Edmonds, J.M. 1978. Numerical taxonomic studies on *Solanum*L. section *Solanum* (Maurella). *Bot. J. Linn. Soc. 76,* 27-51.

Edmonds, J.M. 1979. Biosystematics of *Solanum*L. Section *Solanum* (Maurella). In The Biology and Taxonomy of the Solanaceae (J.G. Hawkes, R.N. Lester and A.D. Skelding, Eds.). Academic Press, London.Pp. 529- 548.

Edmonds, J.M. 1981. The artificial synthesis of Solanum X Procurrens Leslie (*S. Nigrum*L. X S. Sarrachoides Sendtn). *Watsonia 13,* 203-207.

Edmonds, J.M. 1982. Epidermal hair morphology in Solanum L. Section *Solanum. Bot. J. Linn. Sot. 85,* 153-167.

Edmonds, J.M. 1983. Seed coat structure and development in *Solanum*L. Section *Solanum*(Solanaceae). *Bot. J. Linn. Soc. 87,* 229-246.

Edmonds, J.M. 1984. Pollen morphology of *Solanum*L. Section *Solanum. Bot. J. Linn. Soc. 88*, 237-251.

Edmonds, J.M. (1986). Biosystematics of *Solanum sarrachoides*Sendtn.and *S. Physalifolium.*

Edmonds, J.M., &Glidewell.S.M. 1977. Acrylamide gel electrophoresis of seed proteins from some Solanum (Section Solanum) Species. *Plant Syst. Evol. 127,* 277-291.

Edmonds, J.M. 1972. A synopsis of the taxonomy of *Solanum*Sect.*Solanum* (Maurella) in South America. *Kew Bull.27(1)*, 95-L 14.

GoussetC., Cécile C.K., Mulyab I., Mariskab G., Rotinoc L., Pascale B., Aline S.E & Darasinh S. 2005. *Solanum torvum*, as a useful source of resistance against bacterial and fungal diseases for improvement of eggplant (*S. melongena* L.). *Plant Science,* 168(2):319-327.

Harvey, M.H., Mcmillan, M., Morgan, M.R., &Chan, H.W. 1985. Solanidine is present in sera of healthy individuals and in amounts dependent on their dietary potato consumption. *Hum. Toxicol 4 (2),* 187–194.

Heiser, C.B. 1963. Estudio Biosistematico De Solanum (Morella), En El Ecuador. *Ciencia Y Naturaleza 6(2)*, 50-58.

Hellenas, K.E., Nyman, A., Slanina, P., Loof, L., &Gabrielsson, J. 1992. Determination of potato glycoalkaloids and their aglycone in blood serum by high-performance liquid chromatography. Application to pharmacokinetic studies in humans.*J. Chromatogr 573 (1),* 69–78.

Henderson, R.J.F. 1974. *Solanumnigrum*L. (Solanaceae) and related species in Australia. Contributions from the Queensland Herbarium No. *16*:1-78.

Horzog, F., & Gautier-Béguin,D. 2001. Uncultivated plants for human nutrition in Côte d'Ivorie. http://www.fao.org/docrep/w3735e/ w3735e10.htm.12 P.

Jadhav, S.J., Sharma, R.P., &Salunkhe, D.K. 1981. Naturally occurring toxic alkaloids in foods. *Crit. Rev. Toxicol 9 (1),* 21–104.

Jardine, N., & Edmonds, J.M.1974. The use of numerical methods to describe population differentiation. *New Phytol. 73,* 1259-1277.

Jardine, N., & Sibson,R. 1971. Mathematical Taxonomy. John Wiley & Sons, London.

Jennifer, M.E., &James, A.C. 1997. Black nightshades, *SolanumnigrumL* and related species. *International Plant Genetic and Research institute (IPGRI), Italy*.pp. 113.

Kala, C. P. 2005. Ethnomedicinal botany of the Apatani in the Eastern Himalayan region of India. *Journal of Ethnobiology and Ethnomedicine,1(11),* [http://www.ethnobiomed.com/content/1/1/11].

Kusirisin, W., Jaikang, C., Chaiyasut, C., & Narongchai, P. 2009. Effect of polyphenolic compounds from Solanum torvum on plasma lipid peroxidation, superoxide anion and cytochrome P450 2E1 in human liver microsomes. *Med. Chem. 5,* 583–588.

Leslie, A. C. 1978. The occurrence of *Solanumnigrum*.L. X S. sarrachoides send Tn. in Britain. *Watsonia 12(1),* 29-32.

Little, E.L., Woodbury,Jr. R.O., & Wadsworth,F.H. 1974. Trees of puerto rico and the virgin islands. Agriculture Handbook 449.U.S. Department Of Agriculture, Washington, DC. *Vol. 2.*1,024 P.

Mahmood, U., Shukla, Y.N.,& Thakur, R. S. 1983. Non-alkaloidal constituents from *Solanum torvum*leaves. *Phytochemistry 22,* 167–170.

McGehee, D.S., Krasowski, M.D., Fung, D.L., Wilson, B., Gronert, G.A., & Moss, J. 2000. Cholinesterase inhibition by potato glycoalkaloids slows mivacurium metabolism. *Anesthesiology 93 (2),* 510–519.

Mcmillan, M., &Thompson, J.C. 1979. An outbreak of suspected solanine poisoning in schoolboys: Examinations of criteria of Solanine poisoning. *Q.J. Med 48 (190),* 227–243.

Mensinga, T.T., Sips, A.J., Rompelberg, C.J., Van Twillert, K., Meulenbelt, J., Van Den Top, H.J., &Van Egmond, H.P. 2005. Potato glycoalkaloids and adverse effects in humans: an ascending dose study. *Regul. Toxicol. Pharmacol 41 (1),* 66–72.

Mohan M., Bhagat S.J & Sanjay K. 2009. Effect of *Solanum torvum* on blood pressure and metabolic alterations in fructose hypertensive rats. *Journal of Ethnopharmacology,* 126(1): 86-89.

Morris, S.C., Lee, T.H. 1984. The toxicity and teratogenicity of Solanaceae-glycoalkaloids, particularly those of the potato (*Solanumtuberosum*): A Review. *Food Technol. Aust 36 (3),* 118–124.

Morris, T. 1859. Case of poisoning by potatoe-berries. *Br. Med. J 1 (140),* 719–720.

Ndebia, E. J., Kamgang, R., &Nkeh-Chungaganpe, B. N. 2007. Analgesic and anti-inflammatory properties of aqueous extract from leaves of *Solanum torvum*(Solanaceae). *Afr. J. Trad. Complim. Altern. Med. 42:* 240–244.

Nasir, J.Y. 1985. Solanaceae. In: Flora of Pakistan. (Eds.): S.I. Ali and E. Nasir. *Pakistan Agricultural Research council, Islamabad. Fascicle 168,* 1-61.

Nguelefacka,T.B., Mekhfib,H., Dongmoc,A.B., Dimod,T.,Watchoa,P., Johar Zoheire, A., Legssyerb, A., Kamanyi, A.,& Ziyyat, A. 2009. Hypertensive effects of oral administration of the aqueous extract of *Solanum torvum*fruits in l-name treated rats: Evidence from in vivo and in vitro studies. *Journal Of Ethnopharmacology 124,* 592–599.

Null, G. 2001. The biochemical activity of plants.http://www.garynull.com/ documents/phytochemicals/phytochemicals5.htm.15 p.

Orgell, W.H., Vaidya, K. A., & Dahm, P. A. 1958. Inhibition of human plasma cholinesterase in vitro by extracts of solanaceous plants. *Science 128 (3332),* 1136–1137.

Patil, B.C., Sharma, R.P., Salunkhe, D.K., &Salunkhe, K. 1972. Evaluation of solanine toxicity. *Food Cosmet. Toxicol 10 (3),* 395–398.

PIER (pacific islands ecosystems at risk). 2002. Invasive plant species: *Solanum torvum.* Available: http://www.hear.org/pier (accessed: april 22, 2003).

Rahmatullah M., Mukti I.J., Haque A.K., Mollik M.H., Parvin K., Jahan R., Chowdhury M.M & Rahman T. 2009. An ethnobotanical survey and pharmacological evaluation of medicinal plants used by the Garo tribal community living in Netrakona district, Bangladesh.*Advance in Natural and Applied sciences*, 3(3):402-418.

Rao, G.R., & Tandon,S.L. 1969.Relationship between tetraploid *Solanumnigrum*and *Solanumluteum*.*Science and Culture 35*, 688-689.

Rao, G.R., Khan, R.& Khan,A.H. 1971.Structural hybridity and genetic system of *Solanumnigrum*complex.*Bot. Mag. Tokyo 84*, 335-338.

Rao, G.R., Khan, R.& Khan,A.H. 1977.Cytomorphology and nature of sterility barriers of interspecific hybrids of some species of *Solanumnigrum*complex. *Indian J. Genet.& Plant Breed.37(1),* 54-61.

Rajith N.P & Ramachandran V.S. 2010 Ethnomedicines of Kurichyas, Kannur district, Western Ghats, Kerala. *Indian Journal of Natural Products and Resources*2010;1(2): 249-253.

Rayburn, J.R., Friedman, M., &Bantle, J.A. 1995. Synergistic interaction of glycoalkaloids alpha-chaconine and alpha-solanine on developmental toxicity in xenopus embryos. *Food Chem. Toxicol 33 (12),* 1013–1019.

Re R, Pellegrini N, Proteggente A, Pannala A, Yang M &Rice-Evans C. 1999. Antioxidant activity applying an improved ABTS radical cation decolorization assay. Free Radical Bio. Med. 26: 1231–1237.

Seithe, A. 1962. Die haararten der gattung *Solanum*l. und ihre taxonomische verwertung. *Bot. Jahr. 81*, 261-336.

Sivapriya, M., & Srinivas, L. 2007. Isolation and purification of a novel antioxidant protein from the water extract of Sundakai (*Solanum torvum*) seeds. *Food Chemistry104,* 510–517.

Skehan, P., Storeng, R., Scudiero, D., Monks, A., Mcmahon, J., Vistica, D., Warren, J. T., Bokesch, H., Kenney, S., & Boyd, M. R. 1990. New colorimetric cytotoxicity assay for anticancer-drug screening. *Journal of National Cancer Institute 82,* 1107–1112.

Slanina, P. 1990. Solanine (Glycoalkaloids) In Potatoes: Toxicological Evaluation. *Food Chem. Toxicol 28 (11),* 759–761.

Stebbins, G.L. Jr., & PaddockE.F. 1949. The *Solanumnigrum*complex in pacific North America. *Madroño 10,* 70-81.

Symon, D.E. 197). Dioecious Solanums. *Taxon 19(6),* 909-910.

Tandon, S.L.,&Rao,G.R. (1966b). Biosystematics and origin of the indian tetraploid *Solanumnigrum*L. *Curr. Sci. 35(20),* 524-525.

Tandon, S.L., &. Rao,G.R. 1966. Genome analysis in *Solanumnigrum*L. *J. Cytol. Genet. 1,* 41-45.

Tandon, S.L., &. Rao,G.R. 1974. *Solanumnigrum*L. In evolutionary studies in world crops. (Diversity and change in the indian subcontinent) (J. Hutchinson, Ed.). Cambridge University Press, Cambridge, UK.Pp. 109-117.

Tania M.S., Silva R., Jefferson B., Nascimento M., Batista M., Maria F., Agra C &CamaraA. 2007. Brine shrimp bioassay of some species of *Solanum* from Northestern Brazil. *Rev. Bras. Farmacogn*, 17(1):João Pessoa ene./mar.

Tapia A. R., Astudillo A.V.&Uribe R.H. 1996. Resultados preliminaries del efetco de Solanum torvum y plantago major sobre la proliferación de células hematopoyéticas in vivo e in vitro. resumen de ponencias de primer congreso nacional de plantas medicinales de méxico. Tlaxcala, Tlax., *Mexico.* 102- 104.

Udayakumar M., Ayyanar M & SekarT. 2009. Indigenous Knowledge on Medicinal Plants Among the Local People of Puducherry Region (Union Territory), India. *Ethnobotanical Leaflets*, 13: 1401-08.

Venkateswarlu, J., & RaoM.K. 1972. Breeding systems, crossability relationships and isolating mechanisms in the *Solanumnigrum*complex. *Cytologia 37,* 317-326.

Venkateswarlu, J., &Rao,M.K. 1971. Inheritance of fruit colour in the *Solanumnigrum*complex. Proc. *Indian Acad. Sci.,Section B 74,* 137-141.

WiartC., Mogana S., Khalifah S., Mahan M., Ismail S., Buckle M., Narayana A.K & Sulaiman M.2004. Antimicrobial screening of plants used for traditional medicine in the state of Perak, Peninsular Malaysia. *Fitoterapia*,75 (1): 68-73.

Yao, C., Conway, W. S., Ren, R., Smith, D., Ross, G. S., &Sams, C. E. 1999. Gene encoding polygalacturonase inhibitor in apple fruit is developmentally regulated and activated by wounding and fungal infection. *Plant Molecular Biology39,* 1231-1241.

Yuan-Yuan, L.U., Jian-Guang, L. U. O.,&Ling-Yi KONG.2011. Chemical constituents from *Solanum torvum. Chinese Journal of Natural Medicines9(1),* 0030–0032.

INDEX

C

D

E

F

G

H

I

J

K

L

M

N

O

P

Q

R

S

T

U

V

W

Y